*Praise for*
## CAPITALIST PIGS

"Anderson delivers the most thorough account of American pigs ever written, a book packed with fascinating detail on where pigs lived (forests, farmyards, city streets), what they ate (nuts, corn, garbage, the corpses of Civil War soldiers), and how scientists transformed their bodies and their lives to meet the relentless demands of the market. This is the story of how pigs made America, and how America remade the pig."

—Mark Essig,
author of *Lesser Beasts: A Snout-to-Tail History of the Humble Pig*

"J. L. Anderson's *Capitalist Pig*s is a thorough and engaging examination of swine in US agriculture, culture, and history. It will be a standard to judge later histories of Americans' relationships with agricultural livestock and domestic animals."

—Leo Landis,
State Curator, State Historical Society of Iowa
and "the Bacon Professor"

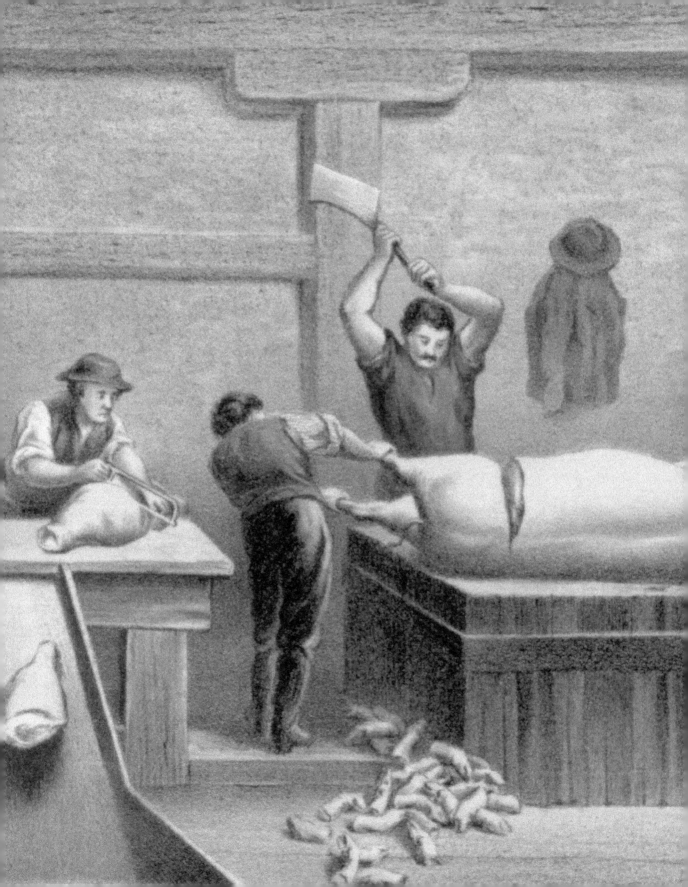

# CAPITALIST PIGS

PIGS, PORK, AND POWER IN AMERICA

## J. L. ANDERSON

WEST VIRGINIA UNIVERSITY PRESS · MORGANTOWN 2019

Copyright © 2019 by West Virginia University Press
All rights reserved
First edition published 2019 by West Virginia University Press
Printed in the United States of America

ISBN:
Cloth 978-1-946684-72-1
Paper 978-1-946684-73-8
Ebook 978-1-946684-74-5

Library of Congress Cataloging-in-Publication Data
Names: Anderson, J. L. (Joseph Leslie), 1966- author.
Title: Capitalist pigs : pigs, pork, and power in America / J. L. Anderson.
Description: First edition. | Morgantown, WV : West Virginia University Press, 2019.
  | Includes bibliographical references and index.
Identifiers: LCCN 2018033302| ISBN 9781946684721 (cloth) | ISBN 9781946684738
  (pbk.) | ISBN 9781946684745 (ebk.)
Subjects: LCSH: Swine--United States--History. | Swine industry--United
  States--History.
Classification: LCC SF395.8.A1 A53 2019 | DDC 338.1/764--dc23
LC record available at https://lccn.loc.gov/2018033302

Book and cover design by Than Saffel / WVU Press
Cover image: This detail of a chromolithograph produced by the Cincinnati Pork Packers' Association for the Vienna Exposition of 1873 depicts pork packing in Cincinnati, circa 1873, from shackling the leg prior to killing to salting. The image depicts an industrial triumph, with overhead conveyors, massive rendering tanks, and an army of laborers employed in transforming pigs into commodities. Source: Library of Congress. Public domain.

*For Sophie, Lucy, and Daisy*

# CONTENTS

|   | | |
|---|---|---|
| | List of Illustrations | ix |
| | Acknowledgments | xi |
| | Introduction | 1 |
| 1. | Making American Gehography | 11 |
| 2. | Hogs at Home on the Range | 35 |
| 3. | Working People's Food | 61 |
| 4. | Pigs and the Urban Slop Bucket | 83 |
| 5. | To Market, to Market | 105 |
| 6. | Swine Plagues | 135 |
| 7. | Making Bacon and White Meat | 152 |
| 8. | Science and the Swineherd | 181 |
| | Coda: The Future of Hogs in America | 217 |
| | Notes | 225 |
| | Index | 277 |

BRISTLING UP
THE
OTA
MINNESOTA
WISCONSIN
MICHIGAN
HYDRO CEPHALUS
NEW YORK
V.T.
N.H.
MAINE
MASS.
CON.
R.I.
CAPE COD
IOWA
RASKA
STATES
ILLINOIS
INDIANA
OHIO
PENNSYLVANIA
N.J.
DEL
MD.
WEST VIRGINIA
VIRGINIA
MISSISSIPPI R.
KANSAS
MISSOURI
KENTUCKY
NORTH CAROLINA
TENNESSEE
SOUTH CAROLINA
INDIAN TERRITORY
ARKANSAS
MISSISSIPPI
ALABAMA
GEORGIA
ATLANTIC BRINE.
LOUISANA
EXAS
FLORIDA
SPANISH LEGACY
RIO GRANDE STREAM.
KEPT UNDER OR
ALTERED.
CUBA

# ILLUSTRATIONS

1. *Ringing the Pig* .................................................. 4
2. Ossabaw hog in Colonial Williamsburg ............... 9
3. "Porcineograph" souvenir poster ....................... 13
4. Map of hog population, 1860 ............................. 18
5. Map of hog population, 1900 ............................. 22
6. Map of feral swine distribution .......................... 30
7. Cornfield damaged by feral hogs ....................... 32
8. Feral hogs trapped in New Mexico .................... 32
9. Williamsburg farmer ........................................... 40
10. Hogging down ................................................... 46
11. Sketch of Frantz Grim chasing hogs .................. 48
12. Pigs in yokes ....................................................... 49
13. Relict pig shelter ................................................ 58
14. Sketch of Robert Wilson's hog .......................... 59
15. Black Hawk, a Berkshire boar ........................... 60
16. Salting down pork .............................................. 71
17. *The Bivouac Feast* ............................................ 80
18. *The Soap Locks, or Bowery Boys* ................... 90
19. Operations at Fontana Farms ............................ 96
20. Steam shovel at Fontana Farms ......................... 97
21. Cycle of transmission of trichinosis .................. 99
22. Driving hogs to the Chicago Market ............... 111
23. Loading hogs on a rail car ................................ 112

24. Workers at a Chicago meatpacking plant ... 113
25. Pork packing in Cincinnati ... 116
26. N. K. Fairbanks and Company trade card ... 125
27. Hog feeding practice common in nineteenth century ... 138
28. Extracting blood from a hyperimmunized hog, Hog Cholera Research Station ... 146
29. Collecting blood in a pan, Hog Cholera Research Station ... 146
30. Administering live-virus serum, Hog Cholera Research Station ... 147
31. Lard sculpture at Indiana State Fair ... 161
32. Woman making chitterlings ... 164
33. Grocery store in Detroit ... 165
34. Graph of US meat consumption ... 167
35. Poland China boar ... 168
36. Modern versus old-fashioned hogs ... 169
37. Landrace hog ... 170
38. Pork meatloaf wrapped in bacon ... 178
39. Joseph Harris's ideal form of market swine ... 184
40. Castrating a young boar ... 185
41. Pigs grazing on alfalfa ... 187
42. Duroc Jersey hogs eating soybeans ... 191
43. Rustic hog shelter ... 192
44. Placing partitions in a hog house ... 194
45. A mixed-breed sow farrows in an A-frame structure ... 195
46. Sanitation on the Klafke family farm ... 198
47. Detail of a portable shade structure ... 200
48. Sows chained in a gestation house ... 201
49. Lactating sows and pigs in a stall ... 201
50. Interior view of a farrowing barn ... 201
51. Interior view of a finishing barn ... 203
52. Exterior view of a finishing barn and manure pit ... 204
53. Watering hogs on the Biesecken farm ... 215

# ACKNOWLEDGMENTS

Although most of the work of researching and writing a book is conducted in solitude, it also involves a community of scholars, archivists, librarians, friends, and loved ones who shape the final product, and this book is no exception. It should go without stating, however, that any errors of fact or interpretation in this work are mine alone.

I was fortunate to work with Derek Krissoff and the team at West Virginia University Press. Thank you, Derek, for helping me through the process from peer review to publication. Thanks to Sara Georgi, Than Saffel, Joseph Dahm, Paula Clarke Bain, and Abby Freeland for making a beautiful book. My special thanks go to those anonymous reviewers who helped domesticate the manuscript. Their challenges and encouragement kept me moving forward.

Mount Royal University has been a collegial home for the duration of the research and writing. MRU generously awarded me a sabbatical to prepare a preliminary draft of the manuscript. Humanities Department members David Clemis, Steven Engler, Mark Gardiner, and Michael Hawley reviewed my sabbatical application and made valuable suggestions for improvement. Thanks, too, to my then Department Chair Jennifer Pettit, former Faculty of Arts Dean Jeff Keshen, and members of the Faculty Leave Committee for endorsing the application. The Humanities Department Innovation Fund and the Faculty of Arts Endeavor Fund provided some of the resources for reproduction and use rights for many of the illustrations reproduced here. I am grateful for the work of Margot

Calvert and Wendy Witczak who secured more interlibrary loans than I care to count. Several undergraduate research assistants including Jason Allen, Rachel Dornian, Heath Milo, Michael Phillips, Matthew Pryce, and Kate Schantz helped sift through databases of historical agricultural periodicals.

Numerous archivists and experts provided assistance at key points in the project. Becky Jordan, Brad Kuennen, Laura Sullivan, and Michele Christian at Iowa State University made me feel at home, as always. Simon Elliott, Public Services, UCLA Special Collections directed me to the LA Public Library photo collection. Angie Rieck-Hinz, Wright County Extension, graciously offered her photographs for use in this project. Thanks to Bill Kemp and George Perkins at the McLean County Museum of History. At the USDA, Travis Kocurek of APHIS and James Fosse and Stacy Carlson of the ARS provided access to important images. Thank you to Ben West, Joe Corn, and John Mayer for sharing your expertise. Chris Dyer shared his thoughts on the varieties of English pig keeping, while Virginia DeJohn Anderson, Doug Hurt, Scott Murray, Leo Landis, Todd Price, and Rick Finch read portions of an early version of the manuscript and offered suggestions that improved the final product. Thanks also to Sara Morris and Francesca Bray. Trevor Soderstrum generously provided a copy of a bibliography he prepared many years ago during our time together at Iowa State. Keith Brownsey of MRU made a timely gift of Julian Wiseman's *A History of the British Pig*.

I sorely missed Mark Finlay's guidance on this project. Many years ago I approached Mark about coauthoring a book on pigs. He begged off, stating that he had years of work in front of him on a project on Georgia's barrier islands. Mark did, however, read and critique my book proposal and subsequently forwarded copies of some documents he had gathered from various archives. Unfortunately, Mark died in 2013. I miss his steady presence, keen intellect, and generous spirit.

Over the course of my many years in university and public history, numerous teachers and colleagues demonstrated the best of the profession. I want to thank Patrice Berger, the late Ed Homze, Tim Mahoney, Ken Winkle, and Pete Maslowski at the University of Nebraska–Lincoln as well as Ray Hiner, Angel Kwollek-Folland, Don Worster, Bill Tuttle, and especially Rita Napier and Victor Bailey at the University of Kansas. Doug Hurt and Pam Riney-Kehrberg were great mentors at Iowa State.

## ACKNOWLEDGMENTS

Nancy Wente of Living History Farms was one of my best friends and supporters. I did not say thank you often enough when she was alive, but I will always be grateful for her kindness and professionalism.

A special word of thanks is due to Mike Pearlman and Tom Morain, both of whom have profoundly influenced my career. They were both honest about the challenges and rewards of a career in history. I recall that Tom once quipped that a suitable goal would be an epitaph of "Employed Historian." I know just how lucky I am to have found employment I love. Thank you for all of your support and encouragement.

I benefited from an outstanding public education in my hometown of Hastings, Nebraska, and offer a belated thank-you to the faculty and staff at Raymond A. Watson Elementary, Hastings Junior High School, and Hastings Senior High School. Those teachers demonstrated incredible patience, even when students like me sometimes proved resistant to lessons, and they returned the next day ready to do it again.

My family stuck by me through this and every other project. Mom and dad provided all the opportunities and love a person could ever hope for. My grandparents worked hard for their children and grandchildren, showing me so much about farming and family. My in-laws, Mary and John Newman, have taken a keen interest in my work over the years, offering continued encouragement and pleasant diversions. Thank you for your support. To my brother Matt, thank you for critiquing a portion of the manuscript, not to mention everything else you have done to teach me through your exemplary life. Emma, thank you for all the love through the good times and hard times. Being part of us is the best part of my life. There would be no book without you. Our girls have lived with this project long enough to occasionally ask, "Dad, when will the pig book be done?" I am happy to say that it is finished and even happier to dedicate it to you.

INTRODUCTION

# PIGS, PEOPLE, AND POWER IN AMERICAN HISTORY

In contemporary America, hogs are present in every state and raised for local, national, and international markets. Pork and lard occupy grocery shelves and refrigerated cases as well as meat lockers and restaurants. Less apparently, swine by-products are ingredients in an array of products including glue, cosmetics, pharmaceuticals, medical replacement parts, and fertilizer, to name but a few items that occupy our medicine cabinets, cupboards, bodies, and landscapes. To meet the domestic and foreign demand for food and other products, Americans changed the diet, morphology, and environment of the pig, creating a modern, high-tech, multi-billion-dollar industry dedicated to the propagation of approximately 110 million animals for the market every year.

Much of the swinish experience in America bears more than a passing resemblance to that of the human experience. Hogs and humans, once dispersed across the land, now live in more concentrated environments, complete with climate control to enhance comfort. Once subject to seasonal availability of food, people and pigs have a predictable diet free of seasonality. Both species are far more sedentary than previous generations, performing productive "work" in office cubicles or confinements. Rather than view these developments as especially pernicious, however, it is important to recognize that both preindustrial and modern iterations of human and hog society have had costs and benefits. Historic hog farming, while less labor-intensive than that of today, saw a high degree of mortality among young pigs, much like the infant mortality rate among people in early America. Today, most pigs survive and live until their appointment at the slaughterhouse. Along the way, they live lives of comparative ease and comfort, even if we have created new, unanticipated stresses, environmental contamination, and a race to the bottom for many workers and producers in the industry. Of course, it is easy to overstate the similarity between the experiences of people and pigs, but it is notable that we have written our aspirations and desires for ourselves on the species that surround us.

How did we arrive at this point in our relationship with these animals? More specifically, what is the place of swine in American history? We know that swine were a critical part of the farm landscape everywhere Europeans settled, earning the appellation "mortgage lifter" for keeping farms solvent. Pork and lard sustained free and enslaved people and fueled the growth of farms, plantations, canals, railroads, towns, and cities. Salt pork and Spam nourished soldiers and sailors. The by-products of pork production, from bristles to lard, became soap and brushes, candles, lamp oil, and even lubricating oil, all of which were part of American culture and economy. Americans frequently judged swine to be unruly, provoking disputes between neighbors and nations. Hogs were the subject of scientific inquiry and technological transformation. We have repeatedly attempted to shape these animals and their place on the land and our tables. In short, hogs are everywhere in the historical record of America.[1]

For as much as swine occupied physical space, it is important to note they occupied cultural and social space as well. Locating these imagined spaces involves assessing changing tastes and values in addition to landscape. European settlers brought their agricultural and dietary traditions to what they labeled the New World where they encountered Indigenous traditions. The enslaved Africans who survived the Middle Passage had their own traditions, and the mixing of these cultures reshaped not only production but also consumption. In America, then, the place of pigs and pork was varied and dynamic, subject to decisions and actions of the powerful as well as the subjugated.

If the place of pigs and pork in America changed through time, then we face multiple questions. Why were pigs and pork more important to some groups than others? Which parts of the animal were most important for different groups? How did production and consumption patterns change during war? How did governments become involved with solving the problems of farmers and processors, and how did governments attempt to shape dietary habits? What role did hogs play in waste reduction, and why did Americans move hogs out of cities? Why did American farmers move pigs indoors, and why did rural and suburban areas experience a surge in feral hog populations?

The answers to these questions are all about power. Individuals, businesses, and institutions remade the commodities of pigs and pork in diverse settings for production and processing. They expressed control over animals and space by emphasizing efficiency through the application of technology, business organization, marketing, and sustained government intervention. Hog farming is not unique in that sense; humans assert their power over the environment to reshape all kinds of commodities in the marketplace. All of these changes reflected the growth of American capitalism in the centuries since the first Europeans and their pigs arrived in North America and attempted to subdue land and people to their own ends. Americans exercised power over hogs and attempted to transcend limits of production and consumption.

The 1842 oil painting *Ringing the Pig* by William Sidney Mount of Long Island, New York, suggests something about that power relationship and its limits. The man astride the hog uses a pair of pliers to secure a ring through the pig's snout. His right hand holds the pliers that bend the ring through the soft tissue, while his left hand maintains tension on the rope that holds the pig in place and helps him keep his balance. The hog pulls backward in an attempt to escape, constricting the snare on his upper mandible. Mount evoked the sound of the scene with the young boy on the left, covering his ears to mute the high-pitched squeal from the ringed hog, not to mention the agitated grunting of the other pigs in the enclosure. A second boy brandishes a cornstalk to slap at faces or hindquarters to keep the other pigs at bay, allowing the work of ringing to proceed unhindered.

Mount's painting captures the action of a once-common activity used to control animal behavior. Hogs root for their food and without much time and effort can destroy or damage fences, crops, and pastures. Once

**Figure 1**
William Sidney Mount's *Ringing the Pig* suggests the ways in which humans attempted to exert control over hogs. By inserting a ring in the soft tissue of the hog snout, farmers hoped to limit the damage hogs could do to crops, buildings, and fences. Source: William Sidney Mount, *Ringing the Pig* (1842). Gift of Stephen C. Clark. Fenimore Art Museum, Cooperstown, New York. Photograph by Richard Walker. New York State Historical Association. Used with permission.

ringed, however, hogs consume what the farmer provides and are less likely to cause damage. Another means of hog control, a yoke, is against the barn door just to the right of the post. The triangle-shaped wooden yoke placed over the animals' neck prevented a pig from escaping enclosures. The hogs in this painting, set in autumn as indicated by the foliage and the fragments of mature cornstalks and leaves on the ground, were ready for fattening prior to sale or butchering in the coming winter. While

Mount completed *Ringing the Pig* in 1842, it is a window on the ongoing effort by humans to bend the hog to human will, establishing limits for swine even as Americans attempted to secure new land and markets for hogs, lard, and pork in the regional, national, and global economy.

This book is about the complex ways Americans remade hogs and, consequently, either transcended or failed to transcend limits. The history of swine and pork as commodities in America has been a story of repeated attempts to transcend limits that included some remarkable successes and many failures. This is a story of power in the context of farming, urban life, and the processing and marketing of food and other products. It is an exploration of pork and lard as food, the utilization of land and labor, and the costs and benefits of both production and consumption. Americans maximized opportunities for the proliferation of the hog, even amid changing values and ideals about empire, agriculture, cities, and health. Debates over the optimal physical place of swine in our geography, the best ways to care for hogs, and the proper role of pork in the diet all reveal the continued importance of hogs in American life. The work to overcome limits of production and consumption has been a constant source of tension in the human-hog relationship. The thesis of this work is that in our efforts to transcend limits of production and consumption, our successes have contained the seeds of failure.

Several noteworthy studies have reckoned with hogs in historical context. In 1950, Charles Wayland Towne and Edward Norris Wentworth published *Pigs: From Cave to Corn Belt*, an excellent narrative of pigs in America. Nevertheless, it was a book of its time; it assumed that progress was natural and inevitable. Swine as a species and the relationship between hogs and people, the authors implied, were moving toward perfection. Contemporary historians do not share those assumptions. Most recently, Mark Essig and Brett Mizelle each authored concise surveys of pigs in history, addressing global developments and, for Mizelle, representations of pigs in visual culture.[2] In contrast to Towne and Wentworth, Essig and Mizelle were justifiably critical of our contemporary food systems. Mizelle emphasized that the world of pig raising, killing, and processing occurs at a significant geographical and intellectual distance from most of us, which has enabled us to conceive of pigs as products more so than as animals, to the detriment of both species. This insight is important, but neither author provided a detailed examination of changes in livestock husbandry to show what those changes meant for farm people and animals. This project shows the ways in which people have fought over the place of pigs and pork in the area that became the United States. I deal in depth with the work of hog husbandry, animal health, and the use of animals in waste disposal, as well as the place of pork and lard in the economy and the table. This book demonstrates the persistence of English and European tradition in taste and farming and shows how agricultural experts who conceptualized of the pig as machine in the nineteenth century facilitated the development of large-scale and integrated operations of the twentieth.

What is so interesting about the pigs? Describing the animals at the heart of this project is even more important than recognizing previous scholarship. Swine, known taxonomically as *Sus scrofa domesticus*, are a runaway biological success story. First, and most importantly for this study, they are omnivores. Like humans, they can subsist in a wide range of habitat, feeding on seeds, nuts, roots, plants, insects, bivalves, and animal flesh. Hogs reach reproductive maturity at approximately six months of age and may bear two litters per year of up to twelve pigs per litter, depending on diet, living conditions, age, and health of the sow. Gestation is three months, three weeks, and three days. Left alone, sows lactate for several months but will cease soon after weaning. Piglets can consume solid food as early as a week of age, even as they continue nursing.

Adult hogs weigh between two hundred and two hundred fifty pounds but may reach over a thousand pounds in optimal conditions. Smaller hogs can move quickly, with an estimated speed of up to thirty miles per hour over short distances. Swine have an exceptional sense of smell and hearing, which, along with good eyesight, allows them to detect food and threats at great distances. Their canine teeth, if not trimmed at a young age, grow into significant tusks that are useful for digging and defense.

Hogs are intelligent, showing extraordinary problem-solving ability, as anyone who has ever attempted to keep them in enclosures can attest. Furthermore, they are highly social animals that appear to enjoy the company of other creatures, and not just other pigs. The much-discussed cleanliness of pigs is real. True, they root through manure for seeds and consume items we consider waste, but in an enclosure they will deposit feces away from where they eat or sleep. These characteristics have allowed hogs to thrive in most ecosystems and conditions, making them an exceptionally useful animal to settler societies over the past four centuries.

When pigs set hoof on North American soil, they did so as partners in one of the most significant global transformations since the last ice age. This moment, or more accurately group of moments, was the Columbian Exchange—the migration of plants and animals, pathogens and parasites, as well as people and ideas between "Old" and "New" Worlds. European pigs were not the first porcine arrivals in what would become the United States, however. That distinction belongs to animals from Polynesia who arrived in Hawaii sometime around AD 400. These animals traced their ancestry to China, the site of swine domestication between eight and ten thousand years ago. While Native Hawaiians relied on fish for protein, even after the arrival of pigs, chickens, and dogs, hogs quickly became critical to native culture, accounting for a larger portion of the diet by the time of European contact. Captain Cook reported that both domesticated and feral hog populations were present in Hawaii when he arrived in 1778, with pigs used for not only meat but also tribute and sacrifice.[3]

The first recorded swine on continental North America was in 1539, when Hernando de Soto brought Spanish hogs from the island of Hispaniola to Florida. After a winter on Florida's Gulf coast, the initial thirteen hogs had allegedly increased to three hundred by April, although this was likely an exaggeration. The de Soto expedition traveled as far inland as modern North Carolina, Tennessee, Arkansas, and Texas with approximately four hundred pigs to provide meat for his company as well as native people they encountered. When de Soto died in Arkansas in 1542, the herd numbered approximately seven hundred hogs. While some of those animals may have escaped and survived to reproduce a feral population, it is much more likely that native people hunted and killed any escapees. When Pedro Menendez de Aviles and a company of settlers established St. Augustine in Florida in 1565, they shipped hogs from Cuba, although those animals died before they had established a breeding population. Settlers at St. Augustine continued to rely on importations of food from Cuba throughout the time it was a Spanish possession. By the late 1600s, there were reports of wild hogs in the woods near the fort, but it is unclear if the hogs were feral animals or belonged to someone in the settlement.[4]

Hogs also came north overland from Mexico. In the late 1500s, several Spanish squatter expeditions reached the Rio Grande with livestock, including swine, but they soon returned home, their imperial plans thwarted by colonial officials. Juan de Oñate brought pigs northward into New Mexico in 1600, but no successful breeding population emerged, even though the accompanying sheep and goats thrived. Archaeological evidence from Santa Fe confirms the marginal importance of pigs and pork in the region during the colonial period, with pork representing less than one percent of the meat diet. Hogs were present in small numbers on the Spanish missions of the province of Texas after 1675, when Fernando del Bosque brought a drove of pigs north across the Rio Grande. Swine, however, did not rival the importance of cattle, sheep, and goats. A report by the friars in 1801 noted that severe winter conditions, lack of pasture, and predation, especially by wolves, limited their success.[5]

The hogs that accompanied the English, Dutch, Swedish, and German settlers fared much better in terms of establishing a population in the New World. English observers had predicted success for swine in the area first known as Virginia, a claim of land that included a large swath of the east coast of the continent. In *A Direction for Adventurers*, published in 1641, Robert Evelyn described the lower Delaware River region, claiming that nut-bearing trees would produce mast that would feed "Hoggs that would increase exceedingly." While Evelyn sometimes stretched the truth in his attempt to boost settlement in the New World, he was accurate in his prediction about the ability of North American forests to sustain multitudes of pigs.[6]

James Towne, later Jamestown, became home to a variety of domestic livestock, even though it is unknown whether animals were part of the initial voyage. There were at least three sows imported sometime in 1607 or 1608, as John Smith boasted that the population by 1609 had increased from three to "60 and od pigges." Subsequent writers have often taken Smith at his word, but as historian Virginia DeJohn Anderson pointed out, that kind of increase would have required those three sows to have farrowed twice per year with either no or low mortality among the litters, which is difficult to imagine. Even European sows of the period did not farrow more than six pigs per litter, even if they did farrow twice per year. It is much more likely that Smith was papering over Jamestown's numerous problems with an unverifiable report. By 1620, colonists at the settlement reported that there were approximately three hundred hogs along with three hundred cattle, two hundred goats, and eleven horses.[7]

Almost every group of settlers brought livestock, including unspecified numbers of hogs. Although the *Mayflower* in 1620 did not carry any livestock, subsequent arrivals in 1622 and 1623 brought cattle, sheep, goats, swine, and poultry. In 1625, the Dutch at New Netherland brought livestock, described as stallions, mares, bulls, and cows, along with "all the hogs and sheep that they thought expedient to send thither." Adriaen van der Donck reported in 1655 that hogs were "numerous and plenty" in the colony.[8]

**Figure 2**

The first hogs in America were the domesticated offspring of the European wild boar. They were heavily bristled, lean, fast, and well suited to the minimal husbandry practiced by the colonists. Farm experts referred to these hogs with derogatory epithets such as razorbacks and land sharks, but those who dismissed these animals as inferior missed the point: these hogs were survivors. Descendants of the first imported hogs survived in isolation on the Sea Islands of Georgia as feral animals for hundreds of years and are now known as Ossabaw Island hogs. This Ossabaw hog was photographed at Colonial Williamsburg. Source: The Colonial Williamsburg Foundation. Used with permission.

Livestock fared somewhat better in New England and the Mid-Atlantic than along the Chesapeake. This was partly because many human immigrants to New England came as families with an interest in practicing mixed farming rather than the hegemonic tobacco and corn culture of Virginia. The first generation of livestock imports thrived to the extent that colonies developed in the late 1600s often received livestock from previously settled provinces. In 1670, the proprietors of Carolina arranged for the shipment of livestock from Virginia. Hogs became so plentiful in South Carolina that the proprietors subsequently exported hogs to the new colony of Georgia in 1733.[9]

The animals that arrived in the North America were part of a larger, imperial project, one in which the interests of European settlers and their animals increasingly dominated the physical and cultural landscape of the continent. Hogs, in particular, were especially useful in expanding the limits of empire. Due to their omnivorous ways, speed, strength, and prolificacy, they readily adapted to New World conditions with no acclimatization efforts required by the settlers.[10] The exercise of power over people, pigs, and geography was one of the most significant developments in American history, with the fate of empires and nations in the balance.

CHAPTER 1

# MAKING AMERICAN GEHOGRAPHY

In 1876, W. E. Baker of Wellesley, Massachusetts, published a "porcineograph," an illustrated color lithograph of the United States depicting the importance of swine, pork, and lard in the American landscape. The map, a self-styled comic "gehography," included the name and seal of each state alongside a particular dish associated with local or regional cuisine. The message was simple. Swine and pork were omnipresent from coast to coast; ham sandwiches were linked to California, salt pork to Arkansas, scrapple to Pennsylvania, and pickled trotters to Florida. By the 1870s, it was readily apparent to Baker and those who viewed the porcineograph that hogs and pork had been central to the construction of an American empire. This chapter documents the spread of swine across the continent from earliest occupation to the present: the making of American gehography.

Hogs quickly became "agents of empire," as historian Virginia DeJohn Anderson observed, proving valuable allies to European powers as those nations attempted to subjugate and displace native people. Hogs continued to serve American expansionism as the place of pigs shifted from allies to commodities. The Trans-Appalachian West became the site of a massive expansion of Euro-American farming, much of it dedicated to growing corn as feed for cattle and hogs. Even as the United States fractured during the Civil War, hogs played a role in shaping that conflict. Food stability was as important to the Confederacy as gunpowder and lead, and it figured prominently in the conflict over leading hog-producing states such as Kentucky and Tennessee in 1861 and 1862. It continued to dog the breakaway republic as both the US and Confederate armies depleted supplies of southern pigs and pork. The war crippled the southern hog-raising empire, privileging the rival states of the Old Northwest as well as Iowa, Minnesota, and Missouri, creating a new empire for swine that persists to the present.

As capitalist modes of production that emphasized expertise and investment became dominant, the hog-raising empire expanded into areas that had not been traditionally associated with hog farming. These new areas were greenfields, places where there was no buildup of pathogens and parasites in the soil and water due to generations of hog farming. Greenfields enabled hog producers to start fresh and increase the scale of their operations. In the areas where hog farming had once been a significant part of agriculture for small and midsized farmers, feral hogs filled the void. No longer competing with native people for food, the new invaders disrupted commercial farming operations, as well as other businesses, homeowners, and land that had been set aside as wildlife refuges. With fewer people on the land, feral hogs forged their own empire, a product of rural industrialization and the human depopulation of the countryside.

## "FILTHY CUT THROATS": NATIVE ENCOUNTERS WITH SWINE

In 1642, the Narragansett Chief Miantonomi reflected on the tremendous changes wrought by colonists over his lifetime. He noted that his father's generation enjoyed plenty of game and fish, but the Narragansett now faced an uncertain future. "But these English having gotten our land," he observed, "they with scythes cut down the grass, and with axes fell the trees; their cows and horses eat the grass, and their hogs spoil our clam banks, and we shall all be starved." He concluded with a plea for

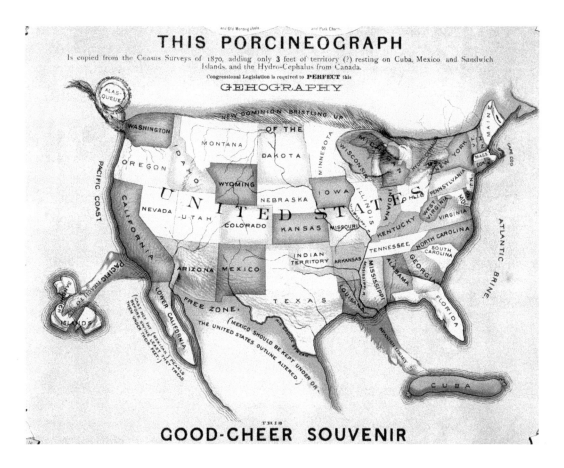

**Figure 3**

The "Porcineograph" souvenir poster of 1876 depicted American "gehography," reflecting the importance of pigs and pork to the American diet and development. Source: Library of Congress. Public domain.

unity among Indigenous groups to drive out the English and restore the old order. It was a justifiable complaint about the English but also swine, their livestock allies. Native Americans were not alone in recognizing the power of swine on the land. Roger Williams of Connecticut described the damage that pigs did to native food sources, writing that they were "most hateful to all natives and they call them filthy cut throats" because they competed for subsistence.[1] Decimated by disease and war, and suffering the constant depredations of swine and other animals, Native resistance to European advances faltered.

A multitude of incidents from every part of the colonies corroborated Miantonomi's understanding of the changes that occurred in the New England landscape. Native people frequently complained about the depredations of European livestock. Hogs ravaged clam beds and grazed on berries and nuts that constituted important food sources for humans. They destroyed reeds, marsh grasses, and small trees that native women used for mats and baskets. Stored food was especially vulnerable. Native farmers used cache pits to store corn and other seed crops underground, but hogs, expert diggers, quickly consumed or spoiled winter supplies. There were few solutions short of rolling logs over the cache pits to prevent rooting, but even that proved imperfect. A 1634 report by natives in Massachusetts that English "swyne" had damaged "Indean barnes of corne" resulted in an extended investigation and court case that was settled by a promise of payment by the colonists. In South Carolina, the deer population declined because European cattle and hogs were competing with deer for food, challenging traditional Yamassee subsistence patterns.[2]

Colonial officials recognized the reality of the situation, but instead of controlling hogs, they urged Aboriginal people to respect European legal custom. The common solution was to construct fences around cultivated land to keep animals out and, if livestock penetrated the fences, to impound the affronting animals. Presumably each animal had a mark such as a brand or ear notch to identify the owner, who would be responsible for paying for damages done by the animal.

Indigenous people understood that livestock were agents of empire. They often responded by hunting the animals that damaged their crops and resources. Archaeologists recovered pig bones at seventeenth-century native sites, indicating that colonists' complaints about native poaching were sometimes accurate. One of the leading historians of colonial-era livestock asserted that hogs were the most common targets of native destruction. This is not surprising, given the prolific nature of hogs and their high degree of adaptability. On the Eastern Shore of the Chesapeake, colonists complained of the "Continuall Trade of killing of hogs."[3] In some cases, there was redress under the law. In 1631, a group of New England colonists complained to a court that a native person had killed an Englishman's pig. The court conducted an investigation and ordered a fine of one beaver skin to cover the loss.[4] Many other charges were false or unproven, but instances of poaching and retaliatory killing of pigs were so common that colonists believed the worst. When residents of Jamestown moved their herd to Hog Island in 1609, they preserved order within the town, but they left the animals vulnerable. During the winter of 1609–1610, known as the Starving

Time, famished colonists could not protect their livestock on Hog Island from natives. During the 1622 Powhatan War, native people attempted to drive the English out of Virginia and attacked not only the settlers but their livestock; "poultry, Hoggs, Cowes, Goats, and Horses whereof they killed great nombers."[5]

Disputes over the repeated trespasses of hogs, cattle, and horses in native fields and gardens led to open warfare. The 1641 Pig War that erupted in New Netherland began after the alleged killing of several pigs by the Raritan people on Staten Island. Colonial official William Kieft used the incident as a pretext for a punitive expedition, which in turn provoked a Raritan attack. While it is not clear who actually killed the pigs, their presence and death exacerbated tensions between native and newcomer.[6] Fifty years of livestock conflicts led to King Philip's War in 1675. The specific trigger for native attacks on the English in 1675 was the fact that colonists had killed three Wampanoag for the murder of a Christian, or "Praying," Wampanoag, who had provided information to the colonists about Indian activities. Throughout summer and autumn of that year the Narragansett and Wampanoag attacked a majority of the New England settlements, killing settlers, taking captives, and destroying crops and livestock. In 1676, colonists and their native allies retaliated, bringing the same kind of "skulking" war to native people, burning villages and targeting crops and livestock.[7]

Some native people selectively adapted European ways, often due to the depletion of game species, decimation by disease, defeat in war, and a hard-won understanding of the ways in which Europeans conceived of land ownership and use. English settlers labeled native hunting ground as waste because it was not "improved" by cultivation and fencing. It was a convenient justification for conquest. The "Praying Indians," those in New England who converted to Christianity in the mid-1600s, accepted livestock husbandry as a means of survival, although there is evidence that native people utilized pigs for food almost immediately after the arrival of Europeans. After defeat in King Philip's War, other native people in the region determined that the adoption of stock raising would allow them to preserve a degree of cultural identity and autonomy in the face of growing English strength.[8]

In the southern Appalachians, the Cherokee people successfully integrated hogs into their agricultural system during the eighteenth century.

While the Cherokee encountered pigs when de Soto moved into their land, it was not until the mid- to late 1700s that observers reported herds belonging to the Cherokee. When John Norton (Teyoninhokovrawen) traveled to Cherokee country in the early 1800s, he noted that hog raising was a widespread practice.[9]

Christian converts and the Cherokee were often successful in livestock husbandry, even as they did not fully meet English expectations for cultural assimilation. In an ironic twist, the hogs that were so numerous and despised by natives for their ubiquity and destructiveness were the European animals most often adopted into native culture. Virginia DeJohn Anderson explained, "What made hogs particularly appealing was their ability to fit niches occupied by animals already familiar to Indians." They were like other game, and their products—not just meat but organs, fat, and hide—were useful like those of other game species. The fact that Aboriginal people adopted what Englishmen associated as the lowest status livestock, far below cattle as a species suited to making proper Englishmen, simultaneously confirmed English prejudice.[10]

Livestock husbandry as an avenue of assimilation proved to be a dead end for many Indigenous people. Many colonists complained that trade in "Indian" pork was detrimental to their economic interests because Native Americans, who raised and sold livestock and meat, undercut prices. In some cases, colonists admonished native people to trade only in game. Sometimes colonists insisted that the hogs native people brought to market needed to have uncut ears to prove that they were not owned by Englishmen. One of the most candid European appraisals was that "Indians ought not to keepe hoggs."[11]

Indigenous people continued to keep cattle, hogs, and sheep, despite the push back from colonists. Pork was already part of native culture by the time of King Philip's War. During Mary Rowlandson's wartime captivity, her native captors offered her fresh pork on at least two occasions.[12] Keeping livestock was sometimes an explicit means of maintaining land claims, as Western Niantic leaders articulated in 1743: "we Could Keep Some Cattle and Sheep and Swine" to refute English assertions that the native people used land only for crops, not livestock.[13] While the colonists were victorious in King Philip's War, hogs and other livestock were also winners. The open range was preserved and the place of swine in the imperial economy was secure.

## THE NEW REPUBLIC: FROM PORKOPOLIS TO THE PIG WAR

After the American Revolution, swine abetted westward expansion into the midcontinent. The same attributes of the pig that enabled European settlers to establish lodgments on the Atlantic and Gulf coasts were useful in the nineteenth century. The predominance of corn in the region facilitated the dependence on swine. As one Indiana newspaper editorialized in 1824, "The principal object pursued at this time, is to raise a crop of corn and a great number of hogs, which embraces almost entirely the whole surplus of the country."[14] A writer for the *American Agriculturalist* asserted in 1845, "It is worth a journey of a thousand miles, any time to take a look at a *small* field of corn of *five hundred acres*" of rich bottom land "and a little herd of *two or three hundred* lusty grunters making away with it on the other." The author argued that eastern farms could not compare in the business of fattening hogs. "Take a trip to the Sciota, the Miami, and the Wabash," he urged, "and one will then get his eyes open and know something about them." Hog raising became so important in the western states that the region became renowned for pigs and pork, so much so that Cincinnati, the nation's leading pork-packing city, became known as "Porkopolis." By the 1850s, hog raising had become so established in the West that a writer from Iowa could claim, "What cotton is to South Carolina, sugar to Louisiana, tobacco to Kentucky, or wheat to Pennsylvania, pork is to Iowa."[15]

In the far western Pacific Coast, hogs brought American and British empires into conflict. The setting was San Juan Island, located in the San Juan de Fuca Straits that separated the United States from what is now known as British Columbia. In 1846, the United States and Great Britain agreed to a border that extended from the forty-ninth parallel into the channel that separated Vancouver Island from the mainland and followed the channel through the straits to the Pacific. Both nations, however, claimed San Juan Island, and both Americans and Britons occupied the island during the 1850s, including a Hudson's Bay Company representative named Charles Griffin, who owned a sheep ranch that was the primary battleground of the 1859 Pig War.[16]

An American settler and an English pig provided the tinder for this conflict. In 1859, American Lyman Cutlar staked a claim on Griffin's ranch, known locally as Belle Vue Farm, where he built a cabin and planted a potato patch. When one of Griffin's hogs entered Cutlar's potato patch and began to forage, Cutlar shot and killed it. Cutlar offered to provide Griffin with a replacement pig or to pay the fair market price for the dead animal, but Griffin insisted on a hundred-dollar payment, far in excess of fair value. American settlers reported the incident to the US Army commander of the department in the Pacific Northwest, who ordered a company of US Infantry to the island to assert American rights. Soon, however,

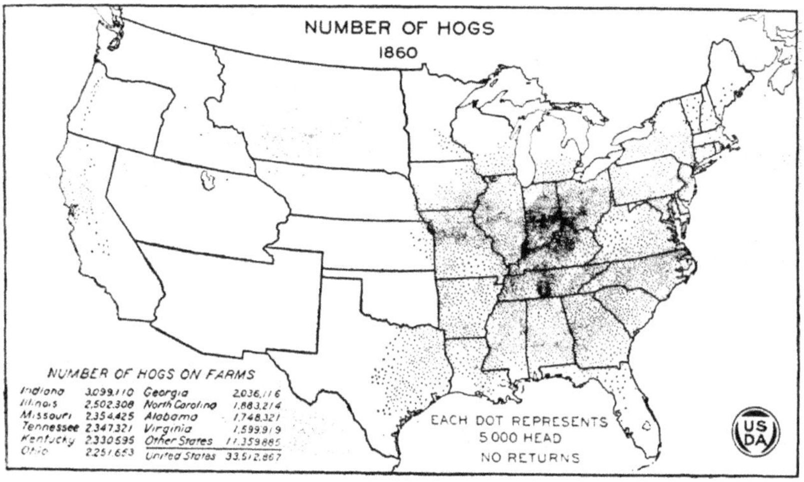

**Figure 4**

The distribution of the hog population in 1860 shows the importance of the West in American hog raising. During the Civil War, the loss of central Tennessee and the fact that Kentucky remained a loyal state meant that the Confederacy was denied access to a large portion of southern pork and lard. Source: USDA. Public domain.

British ships and troops arrived and established camp. A stalemate ensued, but by October cooler heads prevailed. British and American military commanders agreed to a joint occupation, which lasted until 1872. That year an international commission determined that the island belonged to the United States. Subsequent generations downplayed the significance of the Pig War, but it demonstrated the continued place that hogs played on the settlement frontier and the potential they posed for provoking conflict.

## PIGS, PORK, AND THE CIVIL WAR

During the American Civil War, pork was critical for soldiers and civilians in the United States and newly created Confederacy. In 1860, the states that eventually composed the Confederacy were home to seventeen million pigs, while the loyal states possessed nineteen million.[17] Northern and southern farmers alike sold millions of hogs to government buyers or, in the Confederacy, used pork to pay taxes. Pigs were casualties and prizes of war during military campaigns, meeting their end at the hands of foraging parties or driven away from the path of the armies by their owners to protect them for home use or sale. Wartime demand for pork taxed farmers in both nations, resulting in significant declines in livestock numbers during the war years.

Early in 1861, southerners recognized their dependence on outside sources of pork. Historian Paul Gates noted that both the individual Confederate states and the general government imported as much pork as possible from the northwestern states during the secession crisis. Together, the Confederate government, the states, and private individuals in the South purchased approximately 1.2 million of the 3 million American hogs packed during the 1860–1861 season. As long as the war was brief, the Confederacy could survive a meat shortage.[18]

The Confederate commissary likely assumed that pork would occupy a critical place in the soldier's ration, but the geography of southern pork production and the geography of the military campaigns of 1862 in the West dashed those expectations. As noted in chapter 2, Kentucky and Tennessee were the South's leading pig-producing states. Kentucky, however, remained a loyal state thanks in large part to the Lincoln administration's hands-off policy in 1861 and the defeat of Confederate forces at Fort Donelson in February 1862. With the Cumberland River opened to Nashville, the city fell later that month. Kentucky and central Tennessee, the major food-producing region of the state, were behind enemy lines. As the Confederacy's commissary-general of subsistence Lucius B. Northrop wrote, "There is no hope of getting hogs from Kentucky, as parties there feel deserted by the Government" and would sell only for Virginia currency, not Confederate money. Northrup concluded, "So much for hogs."[19]

Despite a Confederate offensive into Kentucky in October 1862 and guerrilla actions in central Tennessee, the US Army maintained control of these states. The cessation of the river trade with the loyal states reduced the flow of pork southward, exacerbating the Confederacy's swine deficit. By 1864, Chattanooga was the base for a US thrust toward Atlanta. Military action, then, shaped the ways in which the Confederate commissary was able to provision its armies.

The presence of US forces on southern soil compromised southern food security at local and national levels. US General Grenville Dodge's 1863 Alabama raid captured 500,000 pounds of southern bacon, while raids into North Carolina in early 1864 accounted for 350,000 pounds. Soldiers of the 3rd Wisconsin Cavalry rounded up a number of hogs "roaming at large" near Balltown, Missouri, in December 1863 and subsequently established a butchering operation that converted southern property into food for federal soldiers.[20] General Philip Sheridan's Valley Campaign of 1864 was notorious for the destruction of barns, granaries, and haystacks, not to mention large numbers of killed and confiscated livestock. In a letter to General Grant, Sheridan estimated that he had taken approximately eleven thousand cattle, twelve thousand sheep, fifteen thousand hogs, and twelve thousand pounds of ham and bacon as well as quantities of grain and forage. While Sheridan may have inflated his numbers to impress his superiors, the Confederacy could not afford even a fraction of such losses. Soldiers and civilians needed meat.[21] In February 1864, one US soldier recorded from rural Hillsboro, Mississippi, that the troops were actively foraging for meat and that there was plenty of "fresh pork" to be had.[22] This "fresh pork," most likely captured on the hoof by the soldiers, was the breeding stock kept over winter for a farm or plantation. While southerners desperately needed the confiscated bacon and hogs that were ready for fattening, the consumption of southern breeding stock compounded the southern food problem for decades to come.

The loyal states also experienced a decline in animal numbers, but at least northerners profited from the sale of livestock rather than losing them through depredations of military forces. Hog numbers in the North declined from nineteen million in 1860 to thirteen million in early 1865, a large share of which became salt pork for the US Army commissary and the export market. For northerners, military purchasing and a surge of exports offset the loss of the southern market. Farmers continued to fear hog cholera throughout the war, with severe losses in the late summer and fall of 1861. For example, Iowa farmers John and Tacy Savage lost over half of their herd to hog cholera in 1863.[23] These losses, however, did not cripple northern agriculture, even in the great western pork-producing region.

## THE PIG IS THE MONARCH OF ALL HE SURVEYS

Few people in 1860 could have envisioned the scale of devastation that would prevail across the South in 1865. The southern livestock business was in shambles. An 1866 US Department of Agriculture census determined that the former Confederate states had experienced a significant loss in the number of animals since 1860. Compared to that year, the South of 1866 retained only 80 percent of sheep, 70 percent of mules, 65 percent of cattle, and 56 percent of swine.[24]

Some southerners even questioned the value of keeping hogs in a post-emancipation world. Before the war, plantation owners feared that enslaved people pilfered pork and pigs, which sometimes resulted in a decision to raise fewer animals. After the war, fears of theft escalated, contributing to a decline in hog numbers. Alabamian James Mallory complained in August 1865 that "demoralization of every kind is the order of the day" and that both whites and freedmen were stealing "cotton, horses, hogs, corn and wheat."[25] In December 1865, John Cunningham of the Anderson District in South Carolina wrote to his uncle, stating that the prospect of hiring freedmen as share tenants looked ruinous. Furthermore, he planned to reduce his livestock holdings. "One thing certain," he stated, "I shall not keep a lot of mules, horses, hogs, cattle etc. with land for a passel of free negroes to support themselves upon."[26] Former slave owners who often praised African Americans as faithful servants in the prewar period (even as they complained about theft) labeled the freedmen as ungrateful takers and thieves during and after the war, eager to get something for nothing.

Not surprisingly, southern recovery was slow. It was only by 1880 that the upper south states surpassed their 1860 hog numbers. In the cotton states recuperation took even longer.[27] In 1877, Thomas P. Janes, Georgia commissioner of agriculture, noted that the state's hog population declined from over 2 million in 1860 to approximately 989,000 in 1870, while the human population had increased, leaving a substantial bacon and pork shortage. Janes wanted farmers to pay more attention to "our important, but sadly neglected friend, *the hog*" to restore prosperity to Georgia farms. He estimated that the difference between home production and purchased pork in 1870 amounted to almost eighteen million in money that could have been kept at home.[28]

The hog became king in the northwestern states, later known as the Midwest. Depressed hog raising in the South was a boon for this corn and hog raising area. The importance of swine in the midwestern economy was not lost on foreign visitors. In an agricultural trade report to the English Parliament in 1880, a correspondent claimed that "no where in the world can such marvelous herds of swine be found as in the corn states of America. . . . Here the pig is monarch of all he surveys."[29]

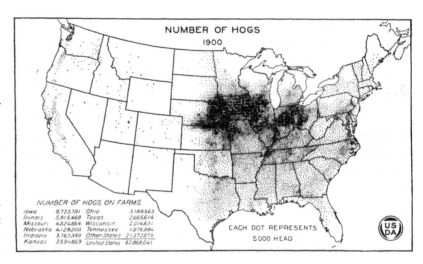

**Figure 5**

By 1900, the Midwest emerged as the leading hog-producing region. Note that hogs are spread evenly across the rest of the eastern half of the country, except for southern Florida, the cutover region of the Great Lakes states, and northern New England. In the West, California's Central Valley and Oregon's Willamette Valley stand out as the region's most important hog-producing areas. Source: USDA. Public domain.

If the pig was a monarch, it owed the crown to cheap grain. As historian Allan Bogue observed, midwestern hog production increased substantially between 1865 and 1900 because hog prices were generally more favorable than those of grain, which made it more efficient to market grain through hogs.[30] A Dutch immigrant farmer in Bon Homme County, South Dakota, explained to his family in 1894, "A farmer has to make his money out of raising hogs. Last year we had to sell our corn, but one does not make nearly as much that way as when one feeds it to the hogs to be sold."[31]

Even as hogs were consistently profitable, there were critics of commercial hog farming. One Iowa farmer acknowledged that hogs were consistent moneymakers, requiring less capital and labor than cattle and horses, but the dedication to hogs and reliance on commercial hog farming was destructive. "If a farmer turns his principal attention to hogs," he stated, "he must have almost his whole farm in corn, from year to year, which will soon exhaust even the rich soil of Iowa."³² After witnessing twenty-five years of farming in Jefferson County, the writer noted that there was an increasing number of exhausted fields in his neighborhood that could not produce good corn. Yet such attention to the environmental costs of hog and corn farming was rare during this period. Presumably, most farmers who relied on hog production could have countered with the argument that the manure from those animals, either carried to the fields or deposited through hogging down, would offset declining fertility.

The changing profile of American immigration boosted hog farming. While immigration from Britain continued to be important, the late 1800s is noteworthy for the surge in immigrants from Germany. These immigrants often favored pork over beef, and in many areas their farming reflected this focus on hogs. In Isanti County, Minnesota, in 1880, German immigrants had over three times as many hogs per farm as Swedish immigrants, and approximately 20 percent more than their American-born neighbors did.³³

The rapid growth of the agricultural Midwest and its promise of security and prosperity through land ownership for immigrants provided a secure footing for hogs. During the first decades of the twentieth century, these north-central states became the home of the vast majority of American hogs. Within the Midwest, Nebraska and Iowa were leaders. During the 1920s the number of hogs per farm ranged from thirty-seven to forty-seven in Iowa and from twenty-eight to thirty-six in Nebraska.³⁴ Throughout the 1930s, Iowa alone accounted for approximately 10 percent of the nation's hog population. In the Midwest, the hog truly was a "monarch of all he surveyed."

## POSTWAR PIGS: CONSOLIDATION, CONTRACTING, AND GREENFIELDS

The geography and structure of American hog production that emerged in the late nineteenth century changed dramatically after World War II. The application of industrial techniques to hog farming and other types of agriculture contributed to a significant restructuring. The mixed farm of the early 1940s rapidly yielded to larger and more specialized operations. Expansive, climate-controlled, specialized buildings, manure-handling facilities such as pits and lagoons, and intensive management characterized new hog farming and required ever-larger amounts of capital and labor. Variations in global and domestic demand for American pork also shaped the industry, not to mention prices for feedstuffs and changes in the meat-packing industry. Local and state regulatory climates helped account for geographic shifts in production, challenging the long-standing dependency between ready access to low-cost corn and hog production. Integration became the norm by the early twenty-first century, as a shrinking number of hog farms produced a growing percentage of American pigs.

The increasing scale of hog production was unmistakable. In 1950, 56 percent of all farms in the United States kept hogs and pigs, but by 1969, only 25 percent reported hogs and pigs. Economist V. James Rhodes described the transition in a 1995 essay, showing that in the 1970s the number of farms producing 200 to 499 hogs per year began to decline, while a decade later the number of producers in the 500 to 999 range started to decline. Meanwhile, the number of producers raising greater numbers of hogs increased. In 1978, one-third of hogs came from farms that marketed more than 1,000 animals per year and 7 percent were from farms with 5,000 head or more. A decade later, those who had 1,000 head marketed 58 percent of total production, and 17 percent came from the 5,000-plus group. Rhodes stated that the 1992 Census of Agriculture indicated that "those percentages rose to 69 percent and 28 percent" for groups with more than 1,000 and 5,000 head.[35]

The trend continued into the twenty-first century. The total number of hog farms declined from 1992 to 2004 by over 70 percent, while the average number of hogs per farm increased from 945 to 4,646 head. Of farms with hogs, 28 percent had over 5,000 head in 1992; in 2004 it was over 50 percent.[36] In the Midwest Corn Belt, those changes were especially dramatic. From 1980 to 2004, Indiana experienced a decline from 24,000 farms with hogs to 3,200.[37] According to a study conducted for the Indiana Department of Agriculture, "the proportion of the state's hog inventory held by operations of less than 1,000 head declined from nearly 45% in 1994 to about 16% in 2004, while farms with over 5,000 head increased in their share of total hog inventory from 19% to 40% over the same period."[38]

Iowa, where hog production had been of prime importance since the mid-nineteenth century, was an extreme case of how hog farming changed. There, the story of the industrialization was characterized by both increasing scale and geographic specialization, even as the state's hog population remained constant. Prior to the 1980s, all ninety-nine counties had significant numbers of pigs, but by the early 2000s, the distribution across the state had narrowed, with the north-central and northwest portions of the state showing rapid growth. While there were only two counties with fewer than 50,000 pigs in 1978, there were twenty-four in 2002. By contrast, there were no counties with more than 400,000 pigs in 1978 but nine counties in 2002, with Hardin and Sioux Counties home to more than 800,000.[39]

What happened to cause such a rapid and thorough change? Farmers experienced shrinking profits in the 1950s, from the late 1970s to the early 1980s, and again in the mid-1990s. Surges in interest rates made it difficult for many farmers to expand, and many of those who borrowed heavily to boost production and construct new facilities were bankrupt when pork prices crashed in the mid-1990s.[40] The years of 1995 and 1996 were especially difficult because hog prices were at twenty-year lows while corn prices reached record highs, making it a losing proposition to feed that grain to hogs. Expansion represented unacceptable costs for those who dropped out, especially if the operator was close to retirement or had health concerns. Furthermore, many children in hog farming families were not interested in carrying on.[41]

Those who could expand often benefited from economies of scale. Even during those difficult years of the mid-1990s, it was less expensive to produce a hundred pounds of hog in a 3,400-sow operation than in one with 650 sows. The largest operations were able to cut costs by three to five dollars per hundredweight over the competition.[42] According to Iowa State University economist John D. Lawrence, the majority of Iowans who quit raising hogs during the 1990s had low investment in facilities. His research indicated that over 87 percent of those who dropped out had open-front buildings with concrete and open lots or pastures. Over half of those who quit had open-front finishing facilities with concrete lots, while only 17 percent had modern confinement operations.[43]

Survivors increasingly turned to contracting, a strategy pioneered in the American South. In the 1930s, feed dealers provided chicks and feed to southern farmers who had taken cotton land out of production as part of New Deal commodity programs. Families with surplus labor and experience with poultry signed contracts to raise chicks provided by the contractor for a set price. The contractor collected the mature birds and processed them, bypassing the intermediary and increasing profits. Pork packers moved rapidly to emulate the model that was so profitable for the poultry industry during the 1940s through the 1960s. In 1992, only 3 percent of US hog farms had production contracts, representing 5 percent of total production. In 2004, the 28 percent of hog farms with production contracts raised 67 percent of the nation's hogs.[44]

Not surprisingly, contract hog farming grew first and fastest in the South. North Carolina emerged as one of the top two hog-producing states in the postwar period. Hogs had always been important in North Carolina, but it was the prevalence of contracting and the abandonment of tobacco farming that provided an opening for new-style hog farming. In 1965, just one year after the US surgeon general declared that there was a link between cigarette smoking and lung cancer, North Carolina established the Swine Development Center to help farmers transition away from tobacco. The center could assist farmers "increase income from swine and other livestock."[45] Like the Georgia cotton farmers of the 1930s who took up contract chicken raising, North Carolina landowners turned to contract hog farming to stay on the land.

One of the leaders of this new style of hog farming in North Carolina was Wendell Murphy. Murphy transitioned from feed mill operator to full-time hog farmer in 1968. That next year, however, hog cholera struck and Murphy endured a complete loss on his three-thousand-head operation. To rebuild, Murphy offered his neighbors, mostly tobacco farmers, contracts to supply them with the equipment, the feed, and eight week-old pigs. The farmer provided land and labor, receiving one dollar for every animal returned to the owner after fifteen weeks of feeding. Murphy then sold the animals to packers. Murphy's contract system spread during the 1970s and 1980s, sparking a surge in hog production throughout the state.[46]

By the late 1980s, Smithfield Foods of Norfolk, Virginia, had relied on farmers in both Virginia and North Carolina to supply its packing facility with hogs, but stricter environmental laws in Virginia prompted the company to move southward. Environmental regulations and organized labor were weak in North Carolina. Furthermore, Murphy had successfully demonstrated the value of the contract system to increase production. Hog production was dispersed across eastern North Carolina in the 1980s, but a decade later the southeastern part of the state had become home to the overwhelming majority of the state's hog population.[47] In 1992, Smithfield opened America's largest pork packing plant in the nation in North Carolina, with a capacity of eight million head per year. In 1993, two economists claimed that North Carolina farmers were "in a rush to put hog production in place to 'fill' this capacity."[48] By the late 1990s, North Carolina rivaled Iowa for the distinction of the leading hog production state in the country.

Meanwhile, midwesterners in the traditional corn-hog belt accepted Wendell Murphy's contracting system. In the mid-1980s, when approximately 50 percent of North Carolina hogs were raised under contract but only 5 percent of midwestern hogs were, Jan and Don Robinette of Ulman, Missouri, accepted a contract from Cargill. The Ulmans had raised hogs but had a hard time making a profit. In 1985, they built a modern finishing barn for fourteen hundred hogs under contract. They explained that people called them traitors. "I'm just trying to keep the farm," Jan stated. "Ownership [of livestock] is what hurt farmers in the first place, so maybe the contract way will help."[49]

Bill Haws, president of National Farms, a large-scale hog producer based in Kansas City, stated in 1984 that the hog business "will inevitably take the same path as the production of chickens. The poultry business has been the birth of extremely cheap food by incredible efficiency of production."[50] The most high-profile example was Premium Standard Farms (PSF), formed in 1988 and purchased by Smithfield in 2007. PSF began operations in north-central Missouri, another sparsely settled area. Kansas and Nebraska also experienced this expansion in the 1990s, with large contract and vertically integrated operations opening in several counties. Even then, many midwestern farmers found contracting to be an unappealing option. In a 1991 poll of over a thousand *Successful Farming* readers in Iowa who raised hogs, only 7 percent indicated that they were actually contract feeders. Those who were not feeding hogs on contract responded negatively to the prospect of doing so, with 55 percent indicating that they would not, 13 percent indicating that they would, and the remainder unsure. When asked if they would quit raising hogs or raise feed on a contract basis, only 20 percent indicated preferring to continue with a contract, while 48 percent claimed that they would rather quit.[51]

For risk-averse farmers, contracting helped even out the market cycle. Farmers who contracted avoided some of the lowest prices but were also not in position to get the best prices. The success of contracting in North Carolina and downward price pressure of the mid-1990s boosted contracting in the Midwest. The growing prevalence of midwestern contracting helped increase production that further depressed prices.[52] Regardless, contracting became a common practice and increasingly accounted for the majority of American market hogs. Contracting and integration also facilitated the expansion of hog operations across the country in areas where it had previously been of minor importance.

Expansion of large-scale hog farming into the Great Plains was a brilliant, if counterintuitive, strategy. The most important advantage was that the Great Plains represented a "greenfield," a region with no history of microbial infection due to the minimal presence of hog farming. This minimized the risk of the kinds of infections that could devastate production in large-scale facilities. Operating expenses were minimal due to low natural gas prices and access to the Ogallala Aquifer for irrigating feed crops and cleaning barns. Labor costs were also low in these remote areas. The sparse population of the Great Plains reduced the complaints about odor and other pollution associated with large-scale hog production. Furthermore, arid conditions reduced the disease threat and facilitated the rapid evaporation of liquid manure.[53]

*Farm Journal* subscribers across the country read about the Great Plains expansion in a 1984 essay titled "The $50 Million Hog Farm." It described the National Farms operation in Holt County, Nebraska, a 16,000-sow, farrow-to-finish facility that would produce over 350,000 animals per year. Thirty acres of manure lagoons held the waste flushed from the buildings. Farrowing and nursery buildings included metal gestational stalls with coated wire floors placed over manure gutters. Just six workers managed 1,300 sows.[54] According to National Farms managers, the facility employed the tried and true rather than the experimental. It was unique because of its size and location rather than any specific on-site practice. National Farms simply followed the market math. In 1990, the cost to produce a hundred pounds of hog was approximately fifty dollars in a 1,600-head operation, while it cost approximately forty-two dollars to make a hundred pounds of hog in an operation with 10,000 animals.[55] Thanks to antibiotics, confinement, manure lagoons, cheap labor and natural resources, and a commitment to year-round farrowing and marketing, hog production entered a new order marked by concentration and integration, expanding in the midcontinent.

During the 1990s, Oklahoma and Texas emerged as leading producers. Seaboard Farms of Merriam, Kansas, opened a new hog processing plant in Guymon, Oklahoma, in 1995, after generous tax subsidies from state and local governments. Texas Farm, a subsidiary of Nippon Meat Packers, opened a major hog production operation in Perryton, Texas, to raise a half million hogs a year for export to Japan, making Ochiltree County the fourth largest hog-producing county in the nation during the late 1990s.[56]

The trend toward increased size and greater integration persisted into the twenty-first century. In 1999, farm journalist Betsy Freese noted that the path forward was clear. For those who wished to use the federal Packers and Stockyards Act to limit the size of hog farms, it was "too little, too late." Freese stated, "The pork industry has turned the corner we've been running toward for five years and sprinting forward for the past year and there may be no going back."[57] In 2002, Congress considered modifications to the act to make it illegal for packers to own, feed, or control (directly or indirectly) livestock for slaughter for more than fourteen days prior to slaughter. The Iowa Pork Producers, which included a larger number of noncontract producers, supported the measure, but it was opposed by the National Pork Producers Council (NPPC) because it did not apply to poultry, which the NPPC claimed would have given poultry an unfair advantage.[58] By 2013, the four largest companies commanded 65 percent of the hog market, with Smithfield alone accounting for 26 percent.[59]

David Meeker, vice president for research and education for the NPPC with a doctorate from Iowa State, explained the rationale driving the industry in the mid-1990s. To stay competitive in a global marketplace, Meeker contended, producers needed to compete at a lower market price, which pressured farmers to find greater efficiencies. "We're doing the right thing, I think, by becoming more efficient and embracing new technology and so forth. They're basically models for the (for lack of a better term) 'North Carolina way of doing business'—big, high tech, and so on."[60] Meeker recognized that there were limits to what was possible in the new regime, but from an industry vantage point survival depended on concentration

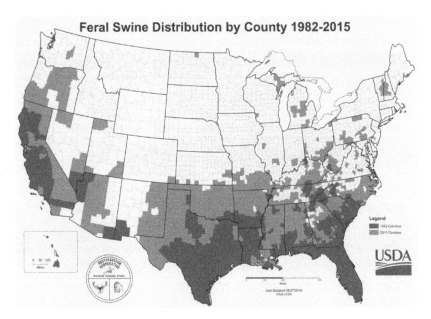

**Figure 6**

Since the 1980s, the range of feral hogs has expanded, most notably northward. There are currently reports of feral hogs in most states and little indication that the feral hog range will contract anytime soon. Source: USDA Southeastern Cooperative Wildlife Disease Study. Public domain.

and integration. Farmers, however, could question the extent to which the new system had worked for them. In 2009, hog farmers received 24.5 percent of the retail value of a hog, half of what they received in 1980.[61]

## HOG WILD: THE PLACE OF FERAL HOGS IN THE RURAL LANDSCAPE

As large-scale, integrated hog production spread into new areas and reshaped hog farming in traditional areas, the number of hogs existing outside of human control increased dramatically. During the 1950s, the feral hog population was relatively contained to the South. The largest populations were present in Florida, southern Alabama and Mississippi, Louisiana, the coastal counties of Georgia and the Carolinas, as well as a handful of counties in southeast Texas.[62] By the 1980s, the range for feral hogs had expanded into nineteen states. In 2014, there were feral populations in thirty-nine states. Stephen Ditchkoff, an Auburn University wildlife ecologist, stated, "Populations in the southeast have exploded. In the Midwest and north it's grown to be a significant problem."[63]

Feral hogs constitute an invasive species threat, uprooting crop fields, pastures, tree plantations, and sod farms. They destroy wildlife habitat and consume endangered plants, prey on amphibian and reptile eggs, and tear up lawns, cemeteries, and golf courses.[64] In Hawaii, they have caused significant damage to rain forests, depleting native species such as fern trees and dispersing the seeds of alien species such as the strawberry guava, which crowds out habitat for rare insect and bird species and encroaches on urban areas.[65] After years of occasional sightings in southern New Jersey, in 2000 groundskeepers at White Oaks Country Club discovered a group of one hundred feral hogs (known as a sounder) tearing up a fairway on the course located in the middle of a 2,675-acre wildlife management area.[66] The USDA currently estimates that several billion dollars of property damage every year is due to feral hogs.

What accounted for the rapid increase in feral hog populations since the 1980s? Some animals were escapees from the farm or the offspring of escapees. Hogs are difficult to contain, and some that escaped never returned. Nevertheless, hogs have escaped since they came to North America, and there have never been so many feral hogs as there are today. Part of the difference is that since World War II, large game hunting has taken on increased importance. Both European wild boar and feral hogs are attractive game animals because of their intelligence and toughness. Landowners looking for extra income could turn hogs loose in large enclosures and then charge hunters for the chance to hunt "wild boar" on their land. This was especially true in the southern states, but the recent expansion into the states that lacked feral hogs suggests that hunting enthusiasts have carried hogs into new areas. Researchers concluded in 2005 that in addition to local escapees of domestic animals, "new populations occurring in areas distant to other feral swine populations are the result of . . . localized but intentional release of domestic swine, or the intentional transport and release of wild-caught or captive feral swine."[67] Bob Butz of *Outdoor Life* explained that the offenders were seldom "farmers or hunting preserve owners who let their fences fall into disrepair." Rather, he claimed, "it is everyday hunters who think it would be fun to have wild hogs running free in the woods close to home."[68]

Ironically, many people viewed hunting as part of the solution to the feral hog explosion, albeit a minor one. States with significant feral hog populations encourage hunting. Laws permitting hunting with dogs and even night hunting were part of the control strategy. As one writer for

**Figure 7**
*Above left*: Feral hogs have become a major presence in the American landscape in the postwar period, causing damage to property and ecosystems. This cornfield has been damaged by feral hogs. Source: USDA Animal Plant and Health Inspection Service. Used with permission.

**Figure 8**
*Above right*: Feral hogs trapped in New Mexico. Trapped hogs are subsequently killed. Source: USDA Animal Plant and Health Inspection Service. Used with permission.

*Outdoor Life* stated, "wild hogs provide year-round hunting opportunities and mighty fine eating."[69] The thrill of stalking these incredibly intelligent and tenacious animals attracts hunters. In some rural areas of the South, hunting hogs is a time-honored practice extending back many generations. In Texas, fine restaurants served feral hog meat, marketed as wild boar, as a lean, healthy, and flavorful game meat. Wild hog meat also appealed to those who rejected industrial pork. As one *Slate* contributor wryly noted, "Invasive pigs are going to be removed only when people decide to take personal responsibility for the problem and go hunting."[70]

Once established in an area, however, feral hogs are impossible to check by hunting alone. Even the writer from *Slate* noted that the current population of hunters would not reduce feral hog numbers. A report published in 2009 confirmed that hunting is not an effective control technique. In fact, in areas where trapping is common, hunting can actually drive hogs away and limit the effectiveness of trapping. Michael Bodenchuk of USDA Wildlife Services stated in a video produced by Mississippi State University, "With over 2 million [feral] hogs in Texas we're not going to barbeque our way out of this problem."[71]

State and federal control experts focused first on lethal techniques to prevent the spread of feral hogs. The most effective lethal approach is to set up a corral trap with bait and then spring the door to capture the offending animals, which are then shot. In places where there has been human activity (including day hunting), night shooting with night-vision scopes and sound-suppressed weapons has been effective. A captured hog with a radio transmitter, known as a Judas hog, is released back into the wild. Once it has reintegrated into the sounder, hunters following the radio signal can locate and kill all the animals.[72]

In early 2017, the Texas Agricultural Commissioner Sid Miller announced that the ideal weapon in the feral pig war was warfarin, a rodenticide. Now prescribed for heart patients as a blood thinner, warfarin can cause death through internal bleeding, sometimes over the course of days or even weeks. Australians used warfarin to reduce feral hog populations before determining that it caused undue suffering. When the commissioner announced that the end of the "'Hog Apocalypse' may finally be on the horizon," opponents countered with concerns about animal welfare as well as the risk to nontarget species. Perhaps most significantly, hunters and trappers feared consuming meat from poisoned hogs. The meat lockers that prepared wild game for hunters won a restraining order in Texas to suspend the use of warfarin in early 2017.[73]

Nonlethal techniques are much less effective in population control. Woven wire hog fence can keep hogs out, but it is expensive to install enough to keep an area hog-free. There is some hope that an injected hormone can serve as a contraceptive, reducing progesterone and testosterone levels and even reducing the weight of ovaries and testes. These chemicals are undergoing testing, however, and the effects are temporary. An integrated control plan is most likely to reduce numbers, but a major obstacle is that much of the range for feral hogs is in private hands, which complicates coordinated control efforts.[74] As long as hunters bring animals into new areas and hogs are hogs, feral swine will continue to multiply and occupy any available ecological niche.

Regular reports of monster feral pigs, often labeled "hogzilla," persist, simultaneously affirming our fears about wild animals and feeding those fears. In 2005, the National Geographic Channel premiered a program titled *Hogzilla*, a documentary about an enormous feral hog that had been killed the previous year on a commercial hunting farm in Berrien County, Georgia. The owner of the hunting operation claimed that the creature was twelve feet in length and weighed approximately a thousand pounds, but the National Geographic team exhumed the carcass and determined that it was closer to eight feet and eight hundred pounds. In subsequent years, numerous claims of new, larger, and more fearsome feral hogs surfaced, igniting debates about whether the latest and greatest hogzilla was truly a feral animal. In at least one high-profile case, the animal in question was a domestic hog that had been hunted and killed just a few days after release. Even though these monster stories have been frequently debunked, they are even more fascinating because of the connection to domestic animals and the descent to wildness.[75]

The reality is that feral hogs, regardless of size, are more challenging than any single hogzilla. Hogs have been effective colonizers in American history due in part to their ability to reproduce quickly and exploit a wide range of habitat. They reach sexual maturity at six months, their gestational period is so short that they can have more than one litter per year, and litters can be as large as a dozen pigs. As omnivores, pigs find food sources wherever they go, including grubs, sea turtle eggs, grasses, and farm crops. Of equal importance is that there are few predators in North America up to the task of taking down a feral hog, even young ones. It is no small irony that as managers of large-scale hog operations implemented ever more stringent bio-security measures to keep domestic hogs alive, the range of feral hogs expanded due to their ability to thrive in multiple ecosystems. These animals pioneered their own greenfields in rural America.

Both pigs and people created American gehography. Hogs served as allies in European conquest and, more recently, enemies on the rapidly growing feral frontier. As highly adaptable omnivores, they often served the interests of capitalist empires. They were central to the exploitation of the western states after the American Revolution and converted the West's prodigious corn crop into a valuable market commodity. The fight over pigs and pork was a significant part of the Civil War, and the result of the war was a consolidation of hog raising in the midwestern states, even as southerners continued to rely on pork. Ultimately, hogs assisted with the recolonization of the Great Plains. The modern gehography succeeded beyond expectations. Unlike the expansion of the late nineteenth and early twentieth centuries, this Great Plains iteration of hog farming was an integrated and concentrated industry. A handful of well-financed corporate directors managed an ever-larger herd, displacing a legion of would-be Jeffersonian agrarians who wanted stability and an independent living on the land. Ironically, with a growing portion of American swine raised in large-scale confinement operations, feral hog populations have risen sharply, exposing the hollowed-out middle of small-scale and midsized farming.

CHAPTER 2

# HOGS AT HOME ON THE RANGE

In October 1861, just three months after the Battle of Bull Run, A. C. Moore of South Carolina toured the battlefield and noted that he witnessed the exposed corpses of "several yankees partially rooted up by hogs." Moore's experiences were not unique. Many Civil War soldiers experienced chilling encounters with hogs and their destructive work. On a memorable night during the Battle of Shiloh in 1862, Augustus Mecklin of the 15th Mississippi reported hearing the sounds of free-ranging hogs among the dead, "quarreling over their carnival feast." A New Jersey chaplain crossed the Bull Run battlefield in late 1862 and stated that dogs and hogs "prowled around their horrible repast" from the second battle at that site. After the Battle of Chickamauga, soldiers built split-rail fence enclosures around the mortally wounded men on Snodgrass Hill to prevent attacks by roving swine. Hogs also consumed freshly amputated arms and legs near the field hospitals. A soldier of the 3rd Wisconsin Cavalry visited a field hospital at Prairie Grove, Arkansas, and reported that "hogs were devouring the limbs that were thrown out of the hospital in a heap for burial."[1]

Twentieth-century writers who described this macabre phenomenon labeled scavenging pigs on the battlefields as wild boar or feral hogs. They suggested that the animals were escapees from enclosure, probably due to the destruction of fences and farmsteads in the armies' path. These writers, however, failed to appreciate the chasm between historical farm practices and contemporary production agriculture. Today, almost every aspect of the pig's life is controlled, but for centuries most hogs were comparatively free of restraint. In this sense, American agriculture functioned extensively rather than intensively. Expansive tracts of land accommodated relatively small numbers of hogs. This system suited American conditions of labor scarcity and ample land. Furthermore, the open range complemented the colonial emphasis on commodities such as tobacco, rice, naval stores, and wheat. Hog raising on the open range suited small-scale farming centered on household production as well as large-scale commodity production. It was an Old World practice that farmers readily accommodated to capitalist production that emerged over the course of the nineteenth century.

During the 1860s, Americans understood this aspect of swine husbandry. Major Edmund B. Whitman, chief quartermaster, US Military District of Tennessee, was in charge of locating unidentified graves in the western theater of the war. When his work began at Shiloh in the spring of 1866, Whitman found human bones scattered across the battlefield. Residents reported that free-range hogs had disturbed the shallow graves and had been "living off the dead" in the woods, fields, and pastures that composed the battlefield. Local farmers asserted that the swine that consumed human remains were tainted, so they refused to round them up for butchering or sale. At a time when calories and cash were in short supply, the southern yeomen farmers of Hardin County, Tennessee, deemed these animals unfit for consumption. By understanding the open range, it is possible to understand their choice and to recognize that the domestic hogs of Shiloh became feral because of the war.[2] The disturbing sights and sounds described by soldiers and civilians were natural consequences of farming practices that extended from the European settlement period into the late nineteenth century.

This chapter presents an examination of hog husbandry from colonization through the American Civil War. It traces the growth of the free range across the continent even as farmers tested the limits of the traditional open range. They experimented with new feeds and feeding regimes, housing, breeding, and sanitary practices. Some reformers urged their fellow Americans to abandon the open range that dominated the landscape. While there were some changes, across much of the nation the open range remained intact.

## MAKING A HOME ON THE RANGE

Descriptions of the open range were ubiquitous among colonial observers. One observer reported that "swine they have in great flocks in the woods."[3] It was a practical solution for colonists who practiced commodity production or were small-scale farmers. At both ends of the production spectrum it was difficult to grow enough food for people, let alone winter feed for livestock, so the range solved a critical problem. In early Maryland, John Lewger, the secretary of the province, counseled Lord Baltimore to purchase hogs to keep at "the head of St. Georges river where all the cheife marshes bee in wch the swine delight."[4] During the 1650s, colonists of New Sweden, in present-day Delaware, allowed their hogs to run along the New Jersey bank of the Delaware River.[5] Farmers performed more remunerative work than tending livestock or raising crops to feed animals that could otherwise feed themselves.

The open range was not unique to the New World. Rather, there were important continuities that Europeans, especially English colonists, understood and practiced. Most important was the role of the forest. According to Joan Thirsk, Britain's leading agricultural historian of the period, "pigs were easily fed in woodlands, some grazing there in all seasons, and having food taken out to them at lean times, or being fattened in the woods on nuts and berries in the autumn."[6] In Britain, the more woodland or marshland present in a given area, the greater the importance of hogs in the local economy. There were places where regular penning was common and farmers used other feed such as barley and peas, but contemporary observers emphasized the central role of woodlands in pig farming. Thomas Fuller described Hampshire County grazing practice of the mid-1600s.

The pigs, Fuller stated, "going out lean, return home fat, without either care or cost of their Owners. Nothing but fulness stinteth their feeding on the Mast falling from the trees, where also they lodge at liberty (not pent up, as in other places, to stacks of Pease), which some assign the reason of the fineness of their flesh." An early eighteenth-century writer described a slight modification in which farmers turned young pigs into the stubble of grain fields after harvest in the summer and then moved them to the forest to fatten on mast. This shared agricultural knowledge and tradition proved useful when hogs arrived on American shores.[7]

It is impossible to understate the importance of range-feeding hogs to the American rural economy, especially for the South. Coastal Chesapeake settlements such as Jamestown quickly exhausted resources, compelling expansion. Middlesex County, Virginia, opened for English settlement in 1648, as downstream planters complained of "over-wrought grounds and the apparent decay of their cattle and hoggs for want of sufficient range."[8] The ever-present desire for land among southern planters, so often associated with the demands of tobacco culture, was also a product of this particular kind of livestock culture. As historian Max Edelson noted, the hoped-for commodities of sugar cane, cotton, ginger, and indigo of Carolina failed during frost and drought, leaving colonists with corn, cattle, and that "weed" species, hogs.[9] As late as 1775, range feeding was still the most common practice on the continent, summarized concisely by a Georgia contributor to *American Husbandry*: "Our swine fare yet better [than cattle], for the woods abound greatly with mast and fruit of various sorts, which they are greedy in finding, and keep themselves fat on," with supplemental food provided in the farmyard in the evening "to induce them to be regular." It was a successful practice, with planters seeing a dramatic increase in their herds, from a few sows to several hundred in a few years' time.[10]

Most farm people relied on the range and provided little restraint for hogs. Robert Cole's three-hundred-acre farm, located in St. Mary's County, Maryland, is an instructive example of how minimalist husbandry made good sense in the seventeenth century. Cole arrived in Maryland in 1652 during a tobacco boom and, like the majority of residents in the Chesapeake, raised it as a cash crop. European demand was strong enough that very quickly tobacco was a means of exchange it its own right, thanks to a shortage of hard currency. In March 1662,

Cole conducted an inventory of his possessions and drafted a will the next month, leaving a remarkable snapshot of a moment in time on a Maryland plantation. When Cole died sometime in 1663, his executors managed the estate on behalf of his seven children for the next ten years. The surviving court records open a window on the lives of southern pigs as well as people.

Cole's hogs and husbandry techniques were common for the region. In 1662, he recorded that twenty nine hogs had "come home," suggesting that they had been abroad in the countryside making their own living for a significant time. This note also indicates that they received at least some feed in the farmyard so that they would identify it as home. In keeping with the minimalist approach, Cole provided no shelter for his hogs, but in 1663 Cole's executor constructed a hog pen.[11] Such minimal care was common. The prominent Virginia planter William Byrd II described the hands-off husbandry practiced by farmers in North Carolina in the 1720s. Byrd reported that cattle and hogs "ramble in the neighbouring marshes and swamps, where they maintain themselves the whole winter long and are not fetched home till the spring."[12]

By 1673, the Cole family herd numbered over fifty head, a respectable increase given that over those twelve years they slaughtered and butchered at least eighty-six hogs and sold another sixty. Maryland farmers could expect between one and ten pigs per litter during this period, although sows farrowed once a year and most commonly delivered three pigs per litter. Litters of four to six pigs were frequent enough to suggest that farmers might hope for six yet also realize that three pigs represented a success.[13]

The lack of care for most livestock, while largely beneficial to the farmer or planter, sometimes had disastrous consequences. Chesapeake farmers endured several harsh winters in the late 1600s that resulted in catastrophic livestock losses. A 1694 report from Maryland indicated that 25,429 cattle and 62,373 hogs died, likely due to exposure, dehydration, and starvation. Such poor conditions also increased their vulnerability to predation. We do not know the total Maryland cattle and hog population for this period, making it difficult to assess the significance of such losses, but the raw numbers plus the fact that the colony conducted a census of weather-related livestock losses suggest that those numbers were disastrous.[14]

**Figure 9**

During the colonial period, farmers relied on hogs to fend for themselves on the open range, bringing them into lots during the fall for fattening on household and farm waste as well as grain, fruit, and root crops. Source: The Colonial Williamsburg Foundation. Used with permission.

Settlers who crossed the Appalachian Mountains after the American Revolution carried the open-range system with them because low-cost husbandry was practical, just as it had been for seventeenth-century colonists. In the 1810s and 1820s, farmers in western New York's Holland Land Purchase grazed their hogs on beechnuts in the woods until butchering time.[15] John Woods, an Englishman who farmed in Illinois in the late 1810s, recorded that while the acorns were thick in 1819, providing ample hog feed, the weather was "too hot and dry for the swine to thrive as well as usual."[16] Another Illinois settler, Gershom Flagg, wrote to his Vermont kin from Edwardsville in 1818, "Hogs will live & get fat in the Woods and Prairies. I have seen some as fat upon Hickorynuts, Acorns, Pecons & Walnuts, as ever I did those that were fatted on corn. All that prevents this country being as full of Wild hogs as Deer," he continued, "is the Wolves which kill the [little] pigs."[17] Matthew Foster's hogs grazed in the woods in Pike County, Indiana. "We have 13 cattle, nearly a hundred hogs," Foster wrote in 1823.[18] Solon Robinson of Indiana visited western Tennessee in 1845 and reported that hogs survived with "little feeding and less care, particularly when 'mast' is plenty, they live[,] move[,] have their being independent of their owner."[19] The West was open-range country.

Texas farmers also practiced open-range farming for hogs. Mary Austin Holley, a cousin of Stephen F. Austin, described conditions in the 1830s: "In many parts of Texas, hogs may be raised in large numbers on the native mast. Acorns, pecans, hickory-nuts &c. with a variety of nutritious grasses and many kinds of roots, afford them ample sustenance during the year."[20] Twenty years later, similar conditions prevailed. A German visitor in Texas reported, "you can keep as many pigs as you wish, and you need not feed them."[21] Occasionally, however, farmers provided supplementary feed. Frederick Law Olmsted noted that Texas hogs could "pick their living from the roots and nuts of the river bottoms," while "a few ears of corn at night brings them all every day to the crib."[22]

Nineteenth-century Americans labeled these open-range hogs as elm peelers, stump suckers, land sharks, alligators, land pikes, and razorbacks. Richard Parkinson stated that such an animal was the "real American hog." Parkinson described them as "long in the leg, narrow on the back, short in the body, flat on the sides, with a long snout, very rough in their hair, in make more like the fish called a perch than anything I can describe." These animals were not only distinctive in appearance. "You may as well think of stopping a crow as those hogs," he continued. "They will go to a distance from a fence, take a run, and leap through the rails three or four feet from the ground, turning themselves sidewise." Given the feast-and-famine nature of the open range, he observed, "These hogs suffer such hardships as no other animal could endure."[23] Indeed, hogs suffered when the mast was light. Scotsman Patrick Shirreff claimed that he saw many pigs running in the forest in Illinois in the early 1830s that he labeled "perfect starvelings."[24]

The open range lasted the longest in the swamps and backcountry. In and around Virginia's Great Dismal Swamp in the first decade of the nineteenth century, local farmers relied on the forest for grazing, where "a great many hogs and some cattle . . . run at large in the woods during summer and keep fat."[25] In South Carolina, traveler James Stuart observed a few years later that hogs "are allowed to roam about during the whole year, feeding on nuts, acorns, &c. which are very abundant in the woods of this country, and occasionally on fallen fruit."[26]

## THE RISE OF FALL FATTENING

During the eighteenth century, a growing number of farmers supplemented mast feeding with grain during the fall. Cornelius Van Tienhoven of New Netherland explained hog husbandry in a letter to prospective Dutch immigrants: "The hogs," he stated, "after having picked up their food for some months in the woods, are crammed with corn in the fall."[27] In 1704–1705, one traveler in Maryland noted that "att the fall of the Leaffe they have fatt Beefe and fatt Porke Comes home to their Doores without giving 'em any Corne when at the same time the people that live upon the River sides and the plantations being thick together they are forced to give there Hogs a great Deall of Corne to ffatt 'em." Fall fattening and over-wintering were costly, though, sometimes amounting to 15 percent of the animal's value.[28]

As colonists cleared ever-larger swaths of forest in settled areas, there was simply less mast available for grazing, leaving planters with few options other than to provide grain. Chesapeake planters, frustrated by low tobacco prices, difficulties in raising a high-quality crop, and soil exhaustion, began to convert tobacco acreage to grain. Nathaniel Burwell, a wealthy Virginia planter, increasingly grew grain on his plantation and used the bran, a byproduct of grain milling, as hog feed in the 1770s. George Washington practiced fall grain feeding on his plantations. In mid-November 1786, Washington's enslaved population began to round up hogs for one month's fattening on corn, with butchering concluded before Christmas.[29]

The open range was less important in the colonies where diversified agriculture was common. Pennsylvanians generally practiced mixed farming, thanks to a liberal policy on religious tolerance that attracted Protestant dissenters who came as family groups. Their farm practices required more enclosures and the collection of manure. Raising multiple varieties of crops and livestock was not conducive to keeping large herds of free-ranging pigs. They enjoyed multiple food sources, including corn, dairy by-products, fruit, and root crops. Still, the open range was common enough in the early days of Pennsylvania, only gradually giving way to pasture and enclosure.

In Pennsylvania, the eclipse of the open range in favor of pasture feeding was apparent by the mid-eighteenth century. In 1748, Swedish visitor Pehr Kalm noted that "great herds" of mast-fed hogs thrived near

Philadelphia. He also observed, "In the immediate vicinity of Philadelphia swine are usually kept in the farmyard," suggesting that at least some Pennsylvania farmers conducted business differently than southerners. Given the larger number of farmers who practiced fall fattening, it is not surprising that the number of hogs in each herd was low compared to those in the southern colonies that received little supplemental feed. Probate inventories from eighteenth-century Pennsylvania indicate that those who owned hogs possessed approximately eight pigs each, much smaller holdings than those of Robert Cole and other Chesapeake planters of the seventeenth century, making it significantly easier to engage in fall feeding with fruit, dairy wash, root crops, and grain. Damaged fruit and the pulp from cider making was good hog feed.[30]

New Jersey farmers also implemented fall feeding practices. In 1756, Joshua Bispham indicated that farmers in his area of southwestern New Jersey often fed a fall fattening ration of two quarts of corn per hog, per day. He selected two hogs weighing seventy pounds and fed them corn and swill, then upped the corn in the ration to gain maximum size and profit just prior to slaughter. That same year, Jacob Spicer complained, "the value of the hogs was totally sunk by feeding them [corn]," given the high price of grain.[31] Abraham Phillips, a tenant farmer, fenced an area as a hog pasture and enclosed stacked oats and corn stalks to keep hogs out. A neighbor claimed the land and took down some of the pasture fence as well as the fences around his stacks. Phillip's "Fattoning Hogs" escaped and over the next few days destroyed much of the winter feed he needed for his high-value cattle and horses.[32]

During the mid-nineteenth century, the use of pasture and feeding hogs "with a little corn" in the fall became increasingly common. Rebecca Burlend, who settled with her family in Pike County, Illinois, during the 1830s, noted that the hogs that ranged during the spring and summer were "scarcely fat enough to be killed," but after "served with a little corn a few weeks . . . thereby become very good bacon."[33] William Cooper Howells of Ohio recalled that while it was possible for hogs to subsist on their own, "it was deemed advisable to pen them up and feed them corn for a few weeks before killing." Howells explained that the farmers kept animals over winter with "a hut of logs, outside the fields, where they would sleep and shelter from storms. Here they were fed and trained to rendezvous, so as to keep them within reach."[34] In a letter from Missouri in 1825, Gottfried Duden stated that the free-range hogs were so successful "that they become quite fat, and do not come home the whole summer long." He noted that when the weather turned, the animals returned to their home place, where the farmers penned them and provided corn prior to butchering. Breeding animals simply remained loose with farmers providing some corn in the farmyard over the winter.[35] In the early 1860s, Iowa farmer Paine Howard of Linn County began to gather his pigs from the range in September, a process that was usually complete sometime in October. In the fall, he fed out his corn, cornhusks, and burned cobs to aid with digestion. He loosed the breeding animals in March, dealing with his hogs in the spring and summer only to remove them from fenced pastures or for spaying and castration.[36]

Hastily built fall fattening pens and shelters were common in all regions of the country. Olmsted described coming upon a Virginia farmstead in the autumn of 1852 and passing the various farmstead dependencies and lots before arriving at the farmhouse. He listed a stable, cattle pen, garden, small cabin, "and a corn-crib and large pen, with a number of fatting hogs in it."[37] Shelters, where they existed, were often simple lean-tos or log buildings. Missouri farmer Stephen Hempstead recorded that slaves on his plantation cut and hauled logs for a hog pen on November 28, 1815. Kentuckian Cassius Clay constructed small, three-sided hog pens made of fence rails covered with boards built up with earth around the three sides.[38] In the 1860s, Paine Howard built a hog shed over the course of nine days one year, work that included seven days to haul logs and lumber and to lay up the structure, one day to lay a floor, and another day to chink the gaps between logs.[39]

Getting animals to the pen was easy in theory but more difficult in practice. Stephen Hempsted of Missouri recorded his experiences of November 23, 1816, in his diary: "I went in the Bottom to hunt my hogs which have not ben up for some time past found *None*." Three days later he went in search of hogs again and "found 15 and got them home and in the pen to fatten *(not got all)*."[40] On one November day in 1845, a Tennessee farmer simply noted, "Hunting hogs all day."[41]

Getting hogs home was easier if there was ample feed. The practice of "hogging down" was a common method of fall fattening in the corn-rich western states that provided a rich diet for swine. Many farmers believed that it was more efficient to let hogs harvest corn rather than investing weeks of labor in the field to harvest it by hand, only to then haul the corn to a crib and subsequently feed it to pigs. An Indiana farm owner on the Upper Wabash River described his system to Henry W. Ellsworth in 1838. He turned hogs into cornfields in mid-September, with about fifteen to twenty hogs per acre, with each field being cleaned up in turn. While this farmer was initially skeptical of hogging down cornfields due to presumed waste, his experiment convinced him that "hogs gather corn in the fields, with little or no waste," if they received regular salt and had water available. The side benefit for this landowner was that his "lands increased in value yearly," without putting "a shovel full of manure on them."[42]

Low-cost, plentiful corn was key for western farmers. A correspondent from Illinois in 1860 described the variety of fattening systems. "In the timber districts the hogs are allowed to run at large," he observed, "and the fences are made hog tight, while on the prairies where fencing material is more expensive, they are confined to pastures of clover and grass, [and] fed a small amount of corn daily." This observation reveals the incremental and variable pace of change as well as the importance of local ecosystems in shaping husbandry practices. Westerners who enclosed animals and used improved breeding stock "turn off fine porkers at an advance over the land pikes," which were fit only for bacon because they were so small that the hams weighed only from eight to fifteen pounds.[43]

Easterners, however, were critical of western husbandry techniques. They agreed that corn was an inexpensive feed in the West, but argued that westerners wasted corn and produced inferior pork. Hogging down was wasteful, they claimed, because the animals trod many ears of corn into the mud and destroyed a significant portion of the crop. The use of a pasture with corn shoveled out of cribs or wagons resulted in the same waste as hogging down and left the animals cold and wet, mired in mud and manure.[44]

Instead of corn, many easterners, especially in New England and New York, used fruit for fall fattening. Farmers with orchards fed sweet apples straight from the tree or boiled and mixed sour apples with cornmeal. Orchards were especially useful if set in a stand of clover or grass where the hogs could not only graze on fruit but also deposit manure and have shade. On one Rensselaer County farm, when the hogs began to manifest a "disrelish for grass, the worm-eaten apples began to fall, sufficiently

**Figure 10**

In the nineteenth century farmers frequently turned hogs into fields of standing corn, a practice known as "hogging down." Rather than invest in the labor-intensive work of bringing in the crop for storage and to feed later, farmers often chose to have animals consume the grain and stalks in the field, all the while depositing their waste directly on the crop land. Hogging down was common in the major hog-producing regions well into the twentieth century. This photograph dates to 1915 or 1916. Source: Ag Illustrated. Used with permission.

matured to become eatable; As they advanced in size and ripeness they became more agreeable, and more nutritious, until the hogs began to fatten rapidly on no other food." As one farmer noted, "Hogs are nothing for corn if they can get apples." Orchards were scarce in the new western states and corn was plentiful, but as soon as farmers established orchards fruit feeding became part of western agriculture. Englishman James Stuart noted that farmers he spoke with in southeast Illinois and southwest Indiana in 1833 fattened hogs on peaches and apples.[45] Orchard feeding required good fences to prevent the animals from destroying the roots and trunks.

## CONTROL: YOKES, RINGS, AND ENCLOSURE

Fall fattening, with either fruit or corn, required controlling the movement of hogs, but achieving control was difficult. Pigs are attracted to plowed soil, green corn, and the opposite sex, not to mention freedom from restraint. They use their proboscis like a shovel to move rocks, dismantle fences, and dig in search of food. Several Philadelphia landowners who drained marshes and meadows were frustrated that free-ranging pigs destroyed their drainage works. In 1703, they petitioned the colonial assembly to pass a law to "Prohibit Swine to Run at large in the said Neck Or Else to Oblige the Owners of them to Ring and Yoke them."[46] James Oglethorpe, Georgia's first colonial governor, ordered the destruction of a group of hogs at Fort Frederica on St. Simons Island after they damaged the earthen fortifications that surrounded the town.[47] The traits that made hogs exceptional survivors made it difficult to control them.

Keeping hogs on islands was an option for those who lived along the coast. These settlers attempted to isolate their pigs and thereby limit any damage to settlers' crops or those of native groups. In 1609, leaders at Jamestown removed the hog herd to Hog Island, just a few miles downstream from the settlement. Kent Island, Maryland, became a stock range in 1629. In the 1630s, New Englanders moved hogs onto several islands in Narragansett Bay, including Prudence, Patience, and Hog islands. A Rhode Island settler rented Prudence Island for three years at a rate of forty pounds sterling and borrowed another forty to purchase the hogs to stock it. Long Beach Island, New Jersey, also served as a stock range. Island animals, by virtue of their isolation, were subject to poaching. Pirates killed several hogs on Sandy Hook, New Jersey, in 1698, but in general the benefits of segregation outweighed the risks.[48]

Colonists used various other control techniques but were never completely successful in keeping hogs in place. Fences or enclosures were the most obvious solutions. It was far easier to fence the cultivated land than to create enclosures, given the modest size of fields that family labor with hand tools could work. Still, farmers needed to split thousands of logs to make stout fences, involving many weeks of labor for preparation and construction. Once built, wooden fences deteriorated quickly, necessitating regular repair and replacement. More importantly, contained hogs required feeding, which meant raising crops to feed animals instead of crops to feed people or the common export staples of tobacco, wheat, or rice.

This meant that it was important to build fences to keep animals out of fields rather than to construct enclosures for animals. As early as 1631, the Virginia assembly decreed, "Every man shall enclose his ground with a sufficient fence," and followed up in 1643 with another law that specified that cleared land should also be fenced. The assembly clarified what constituted a sufficient fence in 1646, defining it as at least four and a half feet tall and "close down to the bottom" to prevent animals from pushing underneath. By 1748, there could be no doubt about what was intended in Virginia: "Lawful fence-a strong sound fence 5 feet high . . . [involving a fence or hedge with ditch and embankment combination] . . . all so close together that horses, mares, cattle, hogs, sheep, and goats cannot creep through."[49]

Hogs, however, regularly penetrated enclosures, compelling legislators to pass new laws specifying liability and damage. In Massachusetts, the General Court decreed that "it shalbe lawfull for any man to kill any swine that comes into his corne." The owner of the offending animal

**Figure 11**

Farmers everywhere in America faced the difficult task of controlling the movement of their hogs. In this early nineteenth-century drawing, artist Lewis Miller of York County, Pennsylvania, sketched Frantz Grim, "a pauper," chasing hogs out of a cornfield. Owners of stray hogs were liable for damages to fenced crops. In the borderlands, stray hogs often provoked conflict between Indigenous people and settlers. Source: York County Heritage Trust. Used with permission.

would then be required to pay damages and could only then claim the carcass. In 1682, East New Jersey passed a similar law that allowed property owners to kill hogs that intruded on their enclosed land or at liberty in town meadows or streets. According to the legislators, "Swine are Creatures that occasion Trouble and Difference amongst Neighbors, and rather prejudicial than beneficial to the Province, while they have Liberty to run at randum, in the Woods or Towns, they being so obstructive to the raising of Corn in the Province, and spoiling the Meadows." The next year, frustrated legislators decided that they had gone too far, rescinding the authorization to kill hogs in meadows or streets and providing for a local option of allowing hogs to range free between October 10 and January 1. In 1785, in the French-held Illinois Country along the Mississippi River, *habitants* agreed to kill pigs found in grain fields and refused indemnification for the pig's owners.⁵⁰

Free-range hogs damaged ecosystems. Hogs in South Carolina competed with deer for forage and favored the roots of young longleaf pine trees. They decimated cane breaks, further displacing native animal and plant species. In New Jersey, a 1789 law set grazing limits and fines for livestock on Peck's Beach and banned swine altogether.⁵¹

**Figure 12**

As early as the seventeenth century, farmers used yokes like these to prevent hogs from escaping enclosures and thereby damaging fences, drainage ditches, pastures, and other improvements on colonial farms and cities. These two hogs in yokes were photographed in the early twentieth century. Source: Ag Illustrated. Used with permission.

In densely settled areas, communities designated particular spaces for hogs or required the use of restraining devices. In Concord, Massachusetts, town leaders dedicated distinct pastures for each kind of livestock. The town's Hog Pen Walk was several hundred acres in size and enclosed with a fence. The close control implied by the named enclosure appears to have been exceptional, however. Even in Concord, community leaders permitted hogs to run at large, so long as owners prevented their animals from damaging fields and meadows.⁵² One solution to controlling the movement of swine was to use a yoke, consisting of a wooden triangle worn snugly around the animal's neck. The pieces of the yoke protruded to prevent animals from squeezing through enclosures or jumping fences. The Massachusetts General Court defined a legal yoke as large as "the full Depth of the Swines Neck above the Neck, and half so much below the Neck; and the sole or bottom of the Yoke Three times as long as the breadth or thickness of the Swines Neck."[53]

Ringing, the practice of inserting a metal ring through the sensitive tissue of the pig snout, was also a common technique to discourage rooting and destruction of fences and other property. Rooting with a ringed snout discomforted the animal, thereby preventing it from digging. The practice did not prevent all the damage hogs could do, especially if they were loose in a grain field where they could graze or trample crops, but it did limit damage. As early as 1640, the town of Plymouth required farmers to ring all swine over the age of three months during the growing season from April through October. Recidivist pigs were to be yoked as well as ringed. Blacksmith John Stewart of Springfield, Massachusetts, regularly ringed all pigs over the age of three months in his community. Stewart traveled through the countryside twice a week to ring hogs during planting season, a time the soil was disturbed and especially attractive to pigs on the loose because they could easily get to the newly planted seeds.[54]

Fencing, yoking, and ringing required establishing ownership of the animal. Marked pigs were easily identified and returned to the owners if stolen, and if hogs damaged property, their owners could be held liable. One or more distinctive notches on either or both ears served as a brand to prove ownership. The town or county captured offending hogs, impounded them, and fined the owner. A 1708 law from New York stated that no swine, "small nor great," would be free to run in the streets, meadows, commons, or in neighbors' fields or otherwise fenced land. The fine was nine pence per animal for the first offense and three shillings for subsequent offenses. If the owner did not claim the animal, recognized by the ear notch, the highest bidder claimed it at auction. Half the proceeds of the sale would go to the person who impounded the pig and the other half to the local government for care of the indigent.[55] It was an imperfect solution to the harm caused by free-ranging swine.

## NO HOME ON THE RANGE?

In densely settled areas, free-range husbandry became increasingly difficult. Tens of thousands of farmers and plantation owners cleared hundreds of thousands of acres while backwoodsmen set fire to the forests to clear land and encourage the growth of understory trees and brush for cattle grazing. Widespread clearing of hardwood forests decreased the availability of nut trees that provided the mast for hogs. Years of mast consumption and grazing on tree roots also damaged hardwood forests. Furthermore, nut-bearing trees were valuable as lumber. Fewer nut trees meant more fast-growing pines and less mast for hogs, notably in the southern piedmont from Virginia through Georgia.[56]

Increased population density of settlers also brought conflict. Historian Robert Jones noted that travelers and diarists in Ohio often remarked on the open-range system through the 1820s, but those comments dwindled in the 1830s, likely due to the fact that Ohio was filling with settlers as well as the growing availability of regular markets for corn-fed hogs.[57] Even so, escaped or trespassing hogs remained a common problem. When Johann Schoepf visited a home in western Pennsylvania, he reported that the husband "was absent looking for his pigs gone astray in the woods."[58] Settlers passed laws much like those of the colonial period that held owners responsible for notching their animals and paying for property damages caused by them. In Ohio, the territorial and state legislatures allowed for the creation of township-level earmark registries, which would prevent the duplication of marks and be useful in recovering stolen or strayed pigs.[59]

Stray hogs plagued Virginian William Walker during the 1830s. Walker, a resident of King and Queen County, endured protracted disputes with his neighbors because his hogs ranged onto Northbank Plantation, owned by Benjamin Pollard and later Pollard's son-in-law, Albert Gallatin Sale. Pollard complained in 1834 that Walker's hogs were "running in my marsh for the last five or six years and have nearly ruined it." The next year, Walker mended his enclosures and even fenced his own section of the riverbank to prevent escapees onto Northbank, but his hogs still found a way out. Sale had not fenced the river-facing side of his cornfield, which made for easy access for Walker's escaped hogs. By the 1840s, Walker calculated that his neighbors, including Pollard and Sale, had killed at least seventy of his wayward hogs. In 1841, Walker won a lawsuit against Sale for killing thirty of his hogs because Sale had not done due diligence by fencing his own field. Walker also complained that his neighbor John Wormley permitted his slaves to kill and butcher Walker's hogs.[60] Presumably, the largest plantation owners who designated one or more enslaved people as swineherds or hog tenders avoided such problems.[61]

Enclosed pasture for hogs was a compromise solution. William Massie, an improvement-minded plantation owner near Charlottesville, Virginia, established fenced lots for intense grazing and rotational cropping, including a ten-acre tract he labeled Hog Lot. Massie hoped that the manure deposited in these tracts would enhance soil fertility and thereby increase crop yields, rescuing the plantation from low productivity and dependence on the vagaries of the tobacco economy. Most planters rejected this kind of "high farming" rotational grazing and cropping system, favoring the short-term returns on cash crops over the long-term investment in soil quality.[62] Even Massie struggled to resist those temptations.

Efforts to close the open range constituted a critical component of improved breeding for pigs and other livestock. As long as farmers fenced their fields to keep animals out, letting livestock shift for their own forage, there could be little control over breeding, not to mention feeding and shelter, all of which were important parts of the new livestock system favored by improvement-minded farmers. Keeping animals in, however, was expensive. A system of enclosed pastures and lots for animals required investments in labor, materials, and future maintenance, and restricted the available food supply for the animals, requiring further labor to haul feed and, in some cases, water.

It was the high cost of fencing that led to the closing of the range in Delaware. The colonial laws of 1739 and 1742 required swine to be ringed or yoked, with any hogs ranging beyond their owner's land subject to capture or death. By the late eighteenth century, the agricultural landscape there resembled that of the rest of the east coast; fields were fenced to keep livestock out. But a growing timber shortage and sandy soil that made it difficult to maintain fences that were "pig tight" led to demands for reform. Opponents of proposed mandatory fencing argued that requiring hogs to be fenced in would be ruinous to the poor, who could barely afford to feed their families, let alone one penned hog. In 1816, the conflict escalated when the town of Smyrna enacted a law that required hog owners to fence their animals. Owners who failed to comply risked having their hogs shot. The state followed suit with a similar law in 1829. Opponents claimed that shooting pigs was a denial of property rights, a claim effectively countered in 1845 when the Superior Court of Delaware declared that a landowner's property line was the equivalent of an invisible fence. A hog that crossed the line was a trespasser and its owner was liable for ensuing damages, representing a shift in property law relating to livestock.[63]

Fence law promoters, however, stressed that significant benefits would follow from enclosing pigs. John Taylor wrote in *Arator* that penning hogs would lessen the destruction of forests and prevent a timber shortage.[64] One of the earliest statements about enclosing stock was in the *Medical and Agricultural Register* in 1806: "The common practice of suffering swine to run at large, during the summer months, is highly injudicious," the writer opined. Damage claims from at large animals were costly and annoying, and furthermore farmers suffered "the total loss of a large quantity of very excellent manure."[65]

A generation later, the complaint was more fully developed. Allowing hogs to run at large or in a pasture was, according to a writer for the *Yankee Farmer*, "bad practice, the hogs 'run away' so much of their flesh that it requires nearly as much to keep them in a thriving state as if they were yarded [penned]."[66] The editor of the *American Agriculturalist* noted in 1845, "We know of no practice more to be reprehended among farmers, than to let their hogs run at large and congregate about their doors when they are fed." He argued that the custom left the "approach to many farm-houses dirty and disgusting in the extreme," and highlighted the financial costs, including expense and trouble of disputes with neighbors like those of William Walker. Furthermore, hogs "glean little abroad, and sadly, waste their flesh in roaming; and what is quite as important to many, they also waste their manure." The solution for this writer was to "keep them up in pens or close fenced fields" or "let swine run in orchards" where they could fatten on grain or fruit.[67]

Agricultural reformers advocated closing the range in many states during the early to mid-nineteenth century. Historian Clarence Danhof asserted that during the 1850s there was widespread agitation for fence laws, citing editorials and reports from Alabama, California, the Carolinas, Indiana, Iowa, Maine, Maryland, Ohio, Virginia, and Wisconsin. In Virginia, many large planters advocated a fence law, but the legislature thwarted them in 1835, when open-range husbandry proved to be more popular. The reformers cited numerous justifications for enclosing livestock, including saving valuable hardwood timber from devastation, capturing prized livestock manure, and making more pork at a faster rate, thereby reducing planter dependence on imported pork. A compromise emerged in which legislatures designated certain rivers as legal fences and allowed communities to exercise a local option for fence laws.[68]

Although few reformers cited the importance of control over breeding stock in their arguments for fence law reform, it ranked among the most important benefits of enclosure. For owners of land sharks and alligator pigs, it made no difference which boar impregnated sows. Most free-ranging pigs were undifferentiated, known only by an adjective, such as "big sow," or by the owner, such as the "Lawson sow."[69] Open-range farmers had minimal contact with these animals. As long as pigs reproduced, all was well. Nevertheless, the investment in a blooded boar or sow necessitated control. A farmer who purchased a valuable Berkshire or Chester White boar hoped to benefit by breeding up his herd, understanding that half of each pig in the litter would reflect the character of the "improved" parent. The boar was simply too valuable to fend for itself or risk impregnating the neighbor's sow without receiving a fee for the service. The proposition was even riskier for purebred sows, which could lose an entire cycle if impregnated by a scrubby land pike or alligator boar. Closing the range was a key to increased profitability for those who were willing to invest more labor in livestock raising.

## COMPLICATING AGRICULTURE: FEEDING, BREEDING, SHELTER

Those who attempted to close the range did so in the spirit of improving agriculture, a project that was a legacy of Enlightenment thinkers who brought science to bear on almost every facet of life. Americans, then, drew on European ideas about livestock husbandry just as they relied on importations of blooded animals from England. The literature of improvement is full of references to English and continental experiences and practices. In "On Feeding Swine" (1821), Pennsylvanian John Linton observed, "I have lately perused with considerable interest as well as satisfaction, a pamphlet compiled by Samuel Parkes of Great Britain on 'The Advantages of using Salt in the various Branches of Agriculture, and in feeding all kinds of farming Stock.'"[70]

One writer for the *New England Farmer* in 1823 reported on the Scottish practice of raising pigs on raw potatoes and concocting a fattening ration of boiled or steamed potatoes with oats, barley, and bean and pea meal.⁷¹ Henry William Ellsworth's *The American Swine Breeder*, published in 1840, is replete with references to experiences and experiments conducted in England, Europe, as well as the United States. In a section on housing, Ellsworth cited not only a plan for a piggery from the *Maine Farmer* but also *Reese's Encyclopedia*, in which the author discussed the Duke of Bedford's English piggery. In the feeding section, Ellsworth cited the Thomas Hale Papers from Oxfordshire in addition to experiments on the value of cooked feed conducted in Scotland, France, Virginia, and Maine.⁷² The rise of improved agriculture in America paralleled that of England and the European continent, evolving together rather than in opposition.

Amid this early nineteenth-century improvement discourse, farm journalists and farmers added new complexity to swine husbandry. Ellsworth stated that "attending to those particulars which are generally deemed unimportant" was necessary for successful hog raising. He cited hog breeder Elias Phinney's caution that "on regular and systematic feeding . . . and clean and dry bedding, the success of rearing and fattening swine very much depends."⁷³ By the mid-nineteenth century, reformers used the term "management" to describe the set of practices that involved improved pens and enclosures, breeding, and feeding. Article titles such as "Management of Hogs" (1840), "Management of Swine" (1849), "Management of a Stock of Hogs" (1855), and "The Hog: Management in Iowa" (1858) are but a few examples of the new vocabulary. These writers shared a common concern with economy in livestock raising, namely, profit. As one writer succinctly stated, the best manager was the one who ensured that animals were "carried as rapidly as possible through the chances of life to that point of development at which they afford a return."⁷⁴ An Illinois farmer stated in 1858 that profitable hog farming required "care and system," which included "suitable shelters, yards, water, conveniences for breeding," not to mention the use of purebred hogs.⁷⁵ Management, in short, promised to improve profits.

To fulfill the dreams of improved and more profitable agriculture, farmers needed to keep their improved pigs carefully fed and in secure enclosures, representing a break with the open-range European tradition that settlers brought in the seventeenth and eighteenth centuries. Close attention to feeding was the ticket to more and better pork. As one New Yorker claimed in 1828, pigs with lots of room to exercise "must expend of that oil in which all their riches consists, and which at rest, they would treasure up. It is therefore a great mistake and want of economy, to confine hogs in the stye and leave them a single hour unsatisfied with food."[76] One writer opined in 1830 that hogs "should never know what liberty is; but should be kept close all their lives, and as inactive as possible."[77] The idea of restricting the movement of these pigs was new for Americans, but numerous writers echoed the demand in the coming years. In 1840, Francis Wiggins critiqued the range system, writing that range-fed and fall-fattened hogs required excess grain, especially because the grain was "fed out in an uncrushed and uncooked state," resulting in inefficient use of farm resources.[78]

Northeasterners complained that a ration of corn alone was too costly and urged farmers to use other crops. In New England, the shorter growing season and greater diversity of farm production meant that there was less corn than on southern or western farms. In 1807, "A Maine Farmer" wrote that farmers could substitute other feeds for corn and cut pig-raising costs in half. Weaning pigs on peas in the summer and feeding them boiled potatoes, carrots, and pumpkins, with supplemental cornmeal and a measure of peas and oats, reduced the dependency on corn, which was just one part of mixed farms in the North, where the growing season was short.[79] Feeding root crops, legumes, cucurbitae (pumpkins and melons), and grains other than corn became popular by the 1840s. In 1840, Henry Ellsworth noted that root crops were gaining in popularity each year and were "regarded as indispensable auxiliaries to other articles in the nourishment and fattening of swine," not to mention other kinds of stock, and were utilized "by the most successful breeders in our country, especially in the Eastern states." High-food-value crops such as mangelwurzels, parsnips, sugar beets, potatoes, cabbages, and artichokes were considered good hog feed.[80]

Farm publications repeated many reports of soaking and cooking feed by boiling or steaming. As early as 1789, Edward Heston of Pennsylvania prepared boiled potatoes and cornmeal to fatten his hogs.[81] A New Yorker, Ezra L'Hommedieu, followed the example of Dr. Elliot of Connecticut and soaked corn prior to feeding it, which persuaded him that he reduced the amount of feed in the fattening ration 10 percent.[82] According to one Pennsylvania experiment, hogs fed on boiled corn and potatoes increased their rate of gain by a third over those fed uncooked corn and potatoes.[83]

Steaming feed involved more labor than soaking, but one farmer defended the practice, claiming that those who steamed food would be "amply paid for all the trouble." The steamer required little fuel and only the labor of "a small boy" to tend the fire.[84] Francis Wiggins reported that experiments indicated that three bushels of cooked corn was worth five bushels of dry corn for fattening.[85] A Kentucky farmer publicized the results of his feed cooking experiment, revealing that the gains from cooked cornmeal were three times greater than what farmers obtained from dry meal, confirming the value of the investment in labor.[86] Specialized diets, usually steamed, accommodated different stages of the animals' development. Farmers fed nursing sows kitchen slop and wash water, ground grain such as oats and corn, and boiled roots. Rations for suckling pigs included skimmed milk or buttermilk, cornmeal, and boiled roots.[87]

Southern elites began to change their hog production strategies, just as improvement-minded farmers in the Northeast had already done. In 1832, a farmer in Twiggs County, Georgia, let his hogs range until late summer, at which time he turned them in to fields of sweet potatoes and peas, letting them clean up each field in succession. This farmer found that in the fall he needed less corn to fatten the animals fed peas and sweet potatoes than those fattened on corn alone. Similarly, in North Carolina, sweet potatoes and peas were important local feedstuffs. In 1845, a correspondent from South Carolina wrote that he had never put much effort into his hogs but planned to change his ways. He proposed to "pen all of them, since I find that the same amount of ground corn and cob-meal fed to them in their range, will enable me to keep them fat in pens," which had the benefit of capturing the manure to use on cropland.[88] Planter James Mallory of Shelby County, Alabama, altered his feeding strategies in the 1840s. As early as 1846, he raised peas for hog feed, and in February 1853, he recorded that his slaves were breaking land for an oat field that would serve as hog pasture.[89]

While feeding was the most important aspect of maintaining the improved pigs, proper shelter and enclosure was a close second. Writers and promoters decried the lack of shelter or poor-quality housing that most farmers provided for their pigs and proposed numerous solutions in the form of hog houses and lots for efficient management. Plans for the ideal hog barn invariably consisted of a feed room that included space for a cooker, a sleeping pen/apartment, and a cellar or vault for collecting manure. The prescribed pens provided just the right amount of space for the animal. The design published in the *Maine Farmer* included sleeping apartments of thirty-six square feet, feeding apartments of fifty-four square feet, and manure vats of ninety-nine square feet.[90]

**Figure 13**

This 1939 photograph shows a relict pig shelter constructed with local fieldstone on the Phipps farm in Sherborn, Massachusetts. Basic shelters like this one, locally known as William Bull's pigpen, were in use into the nineteenth century, especially in regions where hogs were a minor source of farm income. By the mid-nineteenth century, however, frame housing with windows for light and ventilation and sited for drainage was more common across the country. Source: American Antiquarian Society. Used with permission.

Cleanliness of animals and enclosures was a common element of improved husbandry. "Although they are supposed to be naturally filthy animals," a writer for the *New England Farmer* wrote in 1830, "they thrive better and enjoy better health when allowed clean and airy lodgings."[91] In the *Genesee Farmer*, one writer urged that "every necessary attention should be paid to the comfort, cleanliness, and health of the animal." In 1840, Ellsworth quoted a Pennsylvania farm essayist: "They [hogs] show a disposition to be cleanly, however otherwise it is supposed, and always leave their excrementitious [*sic*] matter in a part of the pen distinct from that in which they lie down. No animal," the writer cautioned, "will thrive unless it be kept clean."[92]

All of the work involved in feeding regimes and the construction of new and improved hog shelters could be justified only if the foundation breeding stock was sound. The descendants of the European wild boar were inexpensive and well suited to the inexpensive husbandry of the open range or sty. But English breeders, inspired by Robert Bakewell and others, developed specific breeds of animals that shared a uniform set of valuable traits, although the degree to which those traits showed up across a population was highly variable. The new breeds such as the Berkshire were the product of mixing Old English swine with Chinese and Siamese hogs as well as Neapolitan pigs, many of which were likely descended from Oriental swine. The Chinese animals were less lean and dog-like than the European boar, displaying a characteristic barrel shape.[93] The contrast between old and new could be seen in a set of scale models of pigs created sometime between 1799 and 1810 by the English painter and sculptor George Garrard. Garrard included two traditional English pigs and one that represented a native English and Siamese cross, which

**Figure 14**

In 1809, Pennsylvania artist Lewis Miller sketched Robert Wilson's six-hundred-pound hog. The heavy bristling along the back and front quarters of the animal indicates the predominance of European porcine genetics in the early nineteenth century, predating the widespread importation of pigs with Asian bloodlines during the early to mid-nineteenth century. Source: York County Heritage Trust. Used with permission.

showed the concave or dished face and the barrel morphology that was characteristic of Oriental animals.[94] This kind of animal, in turn, served as the basis of breed improvement in the United States during the nineteenth century.

Despite new attention to management, including breeding, health, feeding, and housing, most farmers in the United States continued to prize the animals that were hardy. Free-range hog husbandry characterized by minimal human effort was common, especially in the South and newly settled areas of the country, much of it settled by southerners. As one conservative Iowa farmer wrote in 1859, "What we want here is a hog with a nose to him so that ho [sic] can root; legs so that he can climb a hazel bush, and hair on him to keep him from freezing." The writer, self-styled as "Old Seed Corner," wanted an animal that could survive without shelter and outrun dogs, "an active thorough-going hog that can take care of himself."[95] Just like the colonists, "Old Seed Corner" and farmers like him in the mid-nineteenth century valued animals that required little care and feed, allowing them to focus on improving their farms by constructing buildings and fences and clearing or breaking land. Range hogs were a

**Figure 15**
The *American Agriculturalist* published this view of the Berkshire boar Black Hawk in 1845. The Berkshire breed was part of the move toward agricultural improvement, with animals from Chinese bloodlines bred to type standards rather than simply relying on natural increase on the open range. Source: American Agriculturalist, June 6, 1845. Public domain.

source of revenue to pay property taxes, the mortgage, and creditors or to reinvest in the farm operation. The open range meant cheap livestock feed, income, and family food security for generations of settlers.

By the time of the American Civil War, hog husbandry was changing. The protestations of "Old Seed Corner" notwithstanding, farming in much of America was increasingly characterized by more intensive management, specifically attention to feeding and breeding, which necessitated closing the range. Still, open-range husbandry prevailed over much of the West and South, and it is not surprising that most of the land that the Union and Confederate armies fought over was open range. Battles, however, were temporary aberrations in the lives of farm animals, presuming that foraging parties of either army did not kill or carry them off. The fact remained, however, that the domestic hogs that feasted on human and horse flesh on Civil War battlefields truly were at home on the range.

CHAPTER 3

# WORKING PEOPLE'S FOOD

When Elisabeth and Wilhelm Koren emigrated from Norway to Decorah, Iowa, in 1853, they encountered a world where pork was distressingly common, at least to their taste. Although Elisabeth liked her new home, she complained about the food, labeling her new country "this land of pork." Her diary contains many references to salt pork, bacon, and ham. She explained that meals varied "from boiled pork to fried pork, rare to well done, with coffee in addition (milk when we can get it), good bread and butter." On December 23, 1853, she awakened at the home of her host to the smell of fried pork. After performing her morning ablutions, she "sat down to the pork, which curiously enough, tastes just as good to me every time I eat it; that is, morning, noon, and night." The Korens occasionally enjoyed a gift of fish, partridge, or chicken, any of which was "a welcome change from that everlasting salt pork."[1]

Of course westerners ate more than just pork; even Elisabeth Koren had an occasional break. The St. John family, New Yorkers who also settled in northeast Iowa five years after the Korens, did not consume pork at all during their first year. Instead, they ate chicken, duck, pigeon, and, on occasion, beef. On December 9, Mary St. John recorded in her diary that the family butchered two hogs that weighed 206 and 152 pounds. The following day Mary recorded, "Have cut up the pork, cleaned the souse, tried the lard, churned mopped, etc." In 1859, at least, they would enjoy hog meat for breakfast and dinner.[2]

Most Americans of the mid-nineteenth century ate pork and, regardless of social position, viewed it as ideal food for working people. This was an English and European tradition in which beef and mutton, in that order, were preferred over pork. Colonists in America successfully replicated that meat hierarchy of their homelands, but by the mid-nineteenth century rapid westward expansion and the growth of agricultural frontiers into the Old Northwest and the plantation Southwest gave pork a boost. Working people and immigrants, free and enslaved, populated these places and shaped regional foodways. During the Civil War era, pork surpassed mutton in the American diet, but beef remained the national preference, even as pork and lard greased the wheels of American industry by feeding much of the American workforce.

### PORK FOR LABOR: EUROPEAN TASTE

When English women and men settled the continent pork was a common food in England and Europe, despite the Judeo-Christian prejudice that pigs were unclean, malevolent, and gluttonous. Europe of the Early Modern period was a place where people ate less meat than contemporary Americans. Meat consumption of the world the settlers left behind, however, was more variable than that of today, with beef, fish, game, fowl, mutton, and pork appearing on tables with regularity. Salted bacon appears frequently in contemporary descriptions of the English diet, but during the seventeenth and eighteenth centuries pickled pork displaced bacon as the most common form of hog meat. Preserving pork in brine had some advantages over salt-cured bacon, including better flavor and less waste (presumably from spoilage), less labor, and less salt. Salted bacon, however, did not disappear. English settlers used salting troughs for making bacon and salted meats throughout the North American colonies.[3]

The most important factor in determining the kind of meat protein on the table and variability was social position. For example, English harvest workers in the thirteenth century generally received bacon as their protein, but during the fourteenth century mutton and beef gradually displaced bacon at harvest meals. This was part of a larger decline in the importance of pork in England after the thirteenth century. Sheep husbandry increased in importance at the expense of pig farming during that period, and there was more extensive grain production to feed a growing urban population. Accelerated forest clearing to meet the demand for industrial charcoal also meant that there was less of the traditional fall hog feed known as pannage (also known as mast, consisting of fallen nuts), which contributed to the decline of pork relative to mutton and lamb. By the seventeenth century, pork became less important in the overall diet of Britons.[4]

Yet most English still viewed pork as ideal for hardworking people. In 1612, one observer stated that salted bacon was the food of "labouring men," and in 1621 Robert Burton stated that "fat bacon" and "salt gross meat" were ideal for farmers and husbandmen. Conflicting evidence suggests that the importance of each type of meat varied by region. In the South and Midlands of England, bacon and beef regularly appeared in probate inventories. Historian Joan Thirsk suggested that after 1750 pork became firmly associated with the poor in Britain when the popularity of potatoes challenged the importance of bread among the diet of the poor. According to this interpretation, the potato skin boosted the nutritional quality of kitchen slop that enabled the pig housed in the dooryard sty to grow faster and larger.[5]

For Spanish colonists from the mast-feeding areas in southern Spain, pork was especially important. Conspicuous pork consumption was a custom that marked a person as Catholic rather than a "heretical" Muslim or Jew, so much so that proclaiming that one ate pork before the inquisitor was a last-ditch effort to prove Christian bona fides. Pork was so common in Spain that it was identified with Spanish culture. Colonial leaders wanted to promote mutton and pork production and consumption in the New World to preserve Spanish culture and Spanish bodies in the harsh climate of North America. They instructed conquistadores to raise pigs, a directive Hernando de Soto heeded by depositing swine on islands to serve as a captive food source for subsequent voyages. As historian Rebecca

Earle explained, colonial Spaniards were so successful in raising pigs and other livestock in central Mexico that average meat consumption in the colonies far exceeded that of Spain. In Mexico, lard displaced olive oil as the cooking fat of choice because it was prohibitively expensive to import oil and the colonists failed to establish local, large-scale olive production. In the northern provinces of Texas and New Mexico, beef tallow was the most important cooking fat due to the prevalence of cattle culture there.[6]

People in western Africa, the cultural hearth for most of the enslaved people brought to America, were familiar with pork, but it was not a favored meat. European explorers and traders carried hogs to Africa sometime during the sixteenth century and likely established successful breeding populations. Islamic authorities dominated many urban areas in West Africa, however, and prohibited pork consumption, although it is likely that some people ate pork in rural areas. Most meat in West Africa, regardless of source, was consumed in modest quantities, often to provide flavor for starches such as tubers, rice, and millet.[7]

## PORK FOR LABOR: EURO-AMERICAN TASTE

The meeting of people from Europe, Africa, and the Americas resulted in the creation of American foodways. For the most part, these New World food practices represented continuity with European tradition, with one notable exception—barbeque. Spaniards interpreted the word that native people used for this kind of preparation as *barbacòa*, a term that migrated to the English language at least by 1661, when it appeared in an account of life in Jamaica. Native Caribbean people used frameworks of sticks to slow-cook meat, much like the drying racks used by native people on the North American mainland for buffalo, fish, venison, and other game. The technique was readily adapted for swine by placing a gutted and split hog on a grate over coals for slow cooking. Edward Ward described barbeque as cooking in "the Indian fashion" in his 1707 account of colonial Jamaica, *The Barbacue Feast; or, The Three Pigs of Peckham*. Ward explained that the entire gutted and split hog, "Heads, Tails, Pettitoes, and Hoofs on," was dressed with a sauce of green pepper and Madeira wine "plentifully daubed on with a Fox's Tail ty'd onto a stick." A 1732 account prescribed the technique: "Take a large Grid-iron, with two or three Ribs in it, and set it upon a stand of iron, about three Foot and a half high, and upon that, lay your Hog . . . Belly-side downwards." The word "barbeque" quickly acquired multiple meanings: a cooking technique, the actual cooked animal, and a social event centered on consuming the slow-roasted meat.[8]

Barbeque became a popular pork preparation on the American mainland by the eighteenth century. Robert Carter, a wealthy Virginia planter, reported attending numerous barbecues during the mid-1700s. One traveler reported attending one such Virginia barbecue in 1774. "These Barbecues are Hogs," he wrote, "roasted whole" with music and "toddy." The barbeque feast reinforced the familial bonds of the Virginia gentry and affirmed their position at the social, economic, and political summit of society.[9] Barbeque occupied an iconic place in the colonial southern diet, but how did pork compare with beef, mutton, and other meats in the New World? The English generally favored beef and mutton in the mother country, but did American conditions provide an opening for pigs and pork? Furthermore, how did colonists preserve and prepare it?

Scholars who have written about the experiences of seventeenth- and eighteenth-century Americans have generalized about the singular significance of pork in the colonial diet. The scholarly and popular consensus was that pork became the most important meat in North America, especially in the southern colonies. There is truth in that story, but it is also misleading. As food historian Sandra Oliver suggested, the perception of pork as *the* major southern food stemmed from its "ubiquity rather than the actual quantity consumed."[10] It is true that colonists, natives, and enslaved people all ate pork throughout the colonial period, regardless of ethnicity, status, or region. They did so, however, to varying degrees, depending on the period and their socioeconomic status and ethnicity. The English meat hierarchy survived largely intact in the English colonies during the initial years of settlement, even as settlement conditions militated against fully re-creating the ideal English palate.

In the Chesapeake, there was a food trajectory from initial settlement of a Native landscape in the early 1600s to a densely settled tidewater landscape by the mid-1700s. It was a trajectory of wild and diverse to domestic and uniform. Wild game and fish bones are present in greater numbers at sites that date to the first decades of settlement, ranging from white-tailed deer to catfish and box turtles. One estimate based on archaeological evidence is that 40 percent of the meat consumption during the first few decades of settlement was from wild animals, including fish.[11]

A comparison of Maryland estate inventories from 1620 to 1745 shows changes in the kinds of animals and food on hand. In St. Mary's County during the 1640s, far more householders owned pigs than cattle, with approximately 100 percent claiming pigs and only 30 percent with cattle. The number of farmers with sheep was negligible during the 1640s, with early importations most likely suffering severe predation. By the 1660s, the number of households with cattle exceeded those with pigs, and by the 1690s, almost 40 percent of farms had sheep. Even if some of those sheep were primarily for wool, there would have been ample surplus animals available for meat so that English colonists could eat mutton like people back home.[12]

Despite the relative plenty of North America, there were important differences in meat consumption between the haves and the have-nots. At the residences of wealthier planters and urban elites, pork constituted less than 20 percent of the meat consumed. By contrast, on two documented sites occupied by tenant farmers in Virginia, pork accounted for between 36 and 39 percent of meat. A suggestive piece of evidence about social hierarchy and diet emerged from a site near Jamestown dating to 1700. Two excavated wells, one associated with the main planter's house and the other next to an indentured servant or slave residence, show major differences in meat consumption. The free and prosperous inhabitants of the plantation ate mostly beef with occasional servings of game, pork, and mutton. The indentured servants or slaves ate equal portions of beef and pork.[13]

Stories like those told by William Byrd II contributed to the inflated perception of southern pork consumption. Byrd, one of Virginia's leading planters, derided neighboring North Carolinians for what he believed to be an unhealthy preoccupation with pork: "The inhabitants of North Carolina devour so much swine's flesh that it fills them full of gross humors." "The Truth of it is," Byrd explained, "these People live so much upon Swine's flesh, that it don't only encline them to the Yaws [skin lesions caused by bacteria], & consequently to the down fall of their Noses, but makes them likewise extremely hoggish in their Temper, & many of them seem to Grunt rather than Speak in their ordinary conversation." Byrd's assessment reflects his class position at the beefy peak of the food hierarchy. It was easy for aristocratic observers to invoke dietary causes to explain the rough or common behavior of North Carolina's "Porcivorous

Countrymen." For all of his complaints about North Carolinians as unhealthy and lazy due to their excessive pork consumption, however, Byrd also regularly ate pork and bacon. What separated Byrd from the Tar Heels was variety. He likely believed that his consumption of beef, mutton, and poultry in addition to pork checked the deleterious effects of too much hog meat.[14]

Even as Byrd penned his narrative in the early 1700s, meat consumption rates of beef and pork were fairly uniform across the southern colonies and consistent with the English meat hierarchy. Beef composed approximately half of the meat consumed in urban and rural settings, with a range of 40 to 60 percent. Pork consisted of approximately 25 percent of the meat consumed, ranging from 10 to 32 percent, with other meat (mutton, game, and poultry) constituting the balance. The investigations of historian Lorena Walsh and other scholars have overturned traditional stories of the primacy of pork in the colonial southern diet.[15]

The repeated references to the good flavor of southern pork contributed to historians' inflation of the importance of pork in the historical record. According to Reverend Andrew Burnaby, who toured the colonies from 1759 to 1760, "The Virginian pork is said to be superior in flavour to any in the world."[16] It is impossible to know whether Burnaby relied on the testimony of Virginia boosters, believed this prior to arriving in the colonies, or heard such reports on his return. He may have simply repeated the earlier published account by Reverend John Clayton, who reported on his Virginia travels to the Royal Society in 1688. "*Swine*," Clayton asserted, "they have in great abundance, Shoats or Porkrels are their general Food; and I believe as good as any Westphalia, certainly far exceeding our English."[17] Burnaby encountered the ubiquitous Virginia ham. "Even at Williamsburg," Burnaby noted, "it is the custom to have a plate of cold ham upon the table; and there is scarcely a Virginia lady who breakfasts without it."[18] In the early nineteenth century, James Stuart claimed that Virginia ham was "admirable" and that much of the pork he sampled in his travels was "remarkably well cured," but he refused to concede that it was better than the best English hams.[19] Still, the presence of these glowing reports in the written record, repeated over the generations, contributed to southern mythmaking.

The story of meat consumption in New England actually paralleled that of the South during the first years of colonization. Wild game and fish were critical for these settlers, just as they were for Chesapeake settlers. In 1630, Governor John Winthrop wrote, "Though we have not beife and mutton etc: yet (God be praysed) we want them not; . . . heere is foule and fish in great plenty."[20] Within a generation, however, beef, mutton, and pork eclipsed game in importance and became common fare in the North. Probate inventories show the importance of salted beef, but the most common salted meat found in New England cellars and storerooms throughout the colonial period was pork, bacon, or ham. In the mid-1600s, almost 25 percent of inventories included some kind of salted pork and approximately 10 percent included salted beef. The ratio of three to one pork to beef prevailed in inventories up to the 1760s, when the ratio of salted pork to salted beef increased to over four to one, a trend into the nineteenth century. The practice of granting a widow's portion from the estate also shows the primacy of salted pork over beef. Between 1700 and 1800, the proportion of salted pork to salted beef remained steady, with the widow receiving a pound and a half of pork for every pound of beef. For widows who lived out of their own storerooms, then, pork was on the table more often than beef.[21] The inventories do not reflect the significance of fresh meat in the diet, but the preference for salted pork over salted beef is clear.

For most New Englanders, however, beef and mutton remained more important than pork for a longer period. Like in Virginia, the archaeological record in New England shows the prevalence of beef in the northern colonial diet. Salted or pickled pork was the rule only on the frontier or at military posts. In rural Deerfield, the number of transactions in the account books of residents suggests consumption patterns of rough equality between beef and pork. Beef and veal accounted for 42 percent, pork 44 percent, and lamb and mutton 12 percent of transactions. Studies of archaeological evidence from house sites in Portsmouth, New Hampshire, that were occupied during the mid-eighteenth century indicate that beef (including veal) was the most important meat consumed in terms of quantity. The number of excavated bones and extrapolations to actual individual animals indicate that beef constituted between 60 and 95 percent of the available meat supply for those households. Pigs accounted for 5 percent up to 30 percent of the total, while sheep and goats accounted for less than 10 percent. Similarly, other excavations across the region show that pork was a distant second behind beef.[22]

Eighteenth-century Bostonians were most successful in sustaining the English meat hierarchy. Archaeologists excavated more cattle and sheep bones than pig bones at urban colonial sites, with mutton even more important than beef. A comparison with excavations at rural sites indicates a distinction between Bostonians and their rural counterparts. While the total proportion of beef, mutton, and pork was the same, pork was more important than mutton to farm families than it was to urbanites. This could have been a function of the relative price of mutton compared to pork, with higher value mutton going to city markets and lower value, easily preserved pork consumed at home.[23]

In the Mid-Atlantic colonies, pork was of mixed importance in the local diet. Account books from the areas of Pennsylvania heavily settled by Germans indicate that pork transactions exceeded those of beef by a factor of two to one. Storekeepers at the colony's numerous ironworks, tanneries, and mills (saw, hemp, paper, fulling) regularly purchased meat from local farmers for resale to workers. A study of these eighteenth-century store records indicates almost twice as many transactions involving veal and beef (fresh and dried) than pork, bacon, gammons, and hams. Mutton was a distant third. The difference between farm accounts in German farm areas and store accounts in industrial settings may suggest that the industrial stores could accept and liquidate a supply of fresh meat before it spoiled, which would privilege beef and veal over salted bacon and ham. The Dutch of New Amsterdam favored pork in their hotchpot, a baked dish of meat, fat, and tubers. On Shelter Island, New York, a planter shipped pork to Caribbean sugar colonies even as both master and slave relied on the lower value cuts during the late seventeenth century. After New Amsterdam became New York City, however, pork declined in importance relative to mutton, reflecting the growing English population and dietary preference.[24]

Salted bacon and ham and pickled pork were the most common forms of preserved pork in the colonies, just as they were in Europe. During the seventeenth and eighteenth centuries the word "bacon" was a generic term, used to describe almost any cut or kind of meat that was salt cured, with the exception of ham. There are even colonial references to bacon made of bear meat. Salt pork, a mostly fatty cut from the back and sides, differed from bacon in that there was little or no meat. Americans smoked their ham (sometimes known as gammon) and bacon, just like the English, with meat hung in the chimney. The first mention of a designated "smoak house" in North America was in York County, Virginia, in 1716. In subsequent years, square Virginia smokehouses measured eight by eight feet up to sixteen by sixteen feet, with steep-pitched, pyramidal roofs, although more simplistic gable-end structures soon became ubiquitous. Settlers hung meat in the smokehouse rafters and built a low fire in the center of the floor to dry the meat.[25]

Fresh pork was rare compared to bacon. Fresh cuts were available only at butchering time, which ranged from mid-November to early January, depending on location. Cold weather facilitated butchering and slowed deterioration of the meat. In cities, in towns, and at taverns it was possible to have fresh pork through the year. Traveler Sarah Kemble Knight ate a plain meal of cabbage, pork, and corn bread in a Rhode Island tavern in 1704. In late August 1744, Dr. Alexander Hamilton, the peripatetic Scottish traveler, noted that while in Connecticut he encountered a man who had "an appointment to eat some roast pigg with a neighbor."[26] The timing of the pig roast coincided with a brief summer break in the farm cycle. The cereal grains had been harvested and were either threshed or in the barn or stack prior to corn harvest. The pig roast was a welcome treat during a slack time between rounds of haying.[27]

Dr. Alexander Hamilton referenced consuming pork on numerous occasions during his American sojourn. Most of his pork entries mentioned bacon or fat pork. A New England host served bacon, chicken, and veal all at the same meal, reflecting a common eighteenth-century custom of serving several meats at one sitting. One midday July meal consisted of fat pork and peas. Given the July date, the piece of fat meat was scraped from near the bottom of the pork barrel with fresh peas from the garden. At another home where Hamilton asked for a dinner of eggs, the host "asked us if we would have bacon fried with our eggs." He declined, leading the host to believe that he was Jewish. The prospect of eggs without bacon was exceptional enough to suggest that the combination was a commonplace meal in the eighteenth century.[28]

Scrapple was one of the more distinctive preparations of pork in the colonies. German immigrants in the Mid-Atlantic region, mostly in Pennsylvania, New Jersey, and Delaware, brought Old World traditions with them to convert the bits of meat that were too small to preserve or

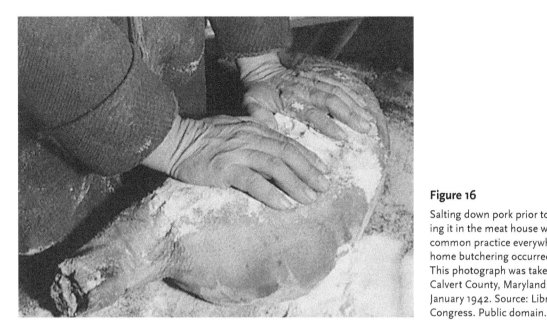

**Figure 16**

Salting down pork prior to hanging it in the meat house was a common practice everywhere home butchering occurred. This photograph was taken in Calvert County, Maryland, in January 1942. Source: Library of Congress. Public domain.

consume on their own, such as trimmings from the head, the organs, and other scraps, into a pudding-like food. At butchering time, these scraps, joints, organs, and the skull were boiled together and the bones removed. Into the broth went cornmeal, buckwheat flour, seasonings, and minced meat, stewed to a stiff consistency. The mixture could then be molded into loaves, cooled, sliced, and then fried or baked within a few days of butchering.[29]

Enslaved people across the colonies consumed a greater percentage of pork in their meat diet than did free people, and they often consumed less valuable parts of the hog than their masters did. When wealthy planters such as William Byrd and Thomas Jefferson ate pork, it was often ham, loin, and some bacon. Elites lived comparatively "high on the hog," the location of some of the most valuable cuts. Enslaved people ate from top to bottom and front to back, but more often from low on the animal—salt pork, organ meat, heads, and trotters. Pig bone fragments that archaeologists recovered from Mulberry Row at Jefferson's Monticello consisted of cranial fragments, lumbar vertebrae, rib bones, and sternums. The stratification of pork consumption was a mirror of an evolving and stratifying colonial society.[30]

## MAKING THE REPUBLIC OF PORKDOM

In the new United States, most English-speaking Americans consumed "at least as much beef as they did pork," according to historian Lorena Walsh, continuing colonial consumption patterns. Walsh cited archaeological evidence from across the country to document the continued importance of beef in American households. Historians overstated pork in the diet, Walsh contended, because it appeared in the probate inventories. But the problem with the probate inventories is that they showed only food on hand, not food actually consumed. Since most Americans consumed beef when it was fresh, it did not appear in the record. Beef was the food of choice through the mid-nineteenth century. In 1854, a writer for *Harper's Weekly* asserted that "the commonest meal in America from coast to coast is steak."[31]

Pork, however, was ascendant, especially in the West and South. Pork displaced mutton in these areas, most likely due to the success of low-cost hog farming in the West. In 1860, Dr. John S. Wilson of Columbus, Georgia, asserted that the United States was the "Hog eating Confederacy, or the Republic of Porkdom." Wilson stated that "so far as meat is concerned, it is fat bacon and pork, fat bacon and pork only, and that continually morning, noon, and night . . . the meat is generally fried, and thus super-saturated with grease in the form of hogs' lard . . . hogs' lard is the very oil that moves the machinery of life." For Wilson, these Americans "would as soon think of dispensing with tea, coffee, or tobacco . . . as with the essence of hog." In the trans-Appalachian West, Americans developed technological innovations in pork packing and, with their hogs, found a mutual dependency among the commodity-producing regions of the South and West and eastern manufacturers. Southerners, most notably enslaved African Americans and poor whites, relied on salted pork, a cheap food source that had also been an important food to laborers in Europe. Beef remained king, but pork was rising.[32]

For most Americans, bacon and salt pork were superior to beef for travel. The architect Benjamin Henry Latrobe made a boat trip up Virginia's Appomattox River in 1796 with "plenty of good ham, bacon, Indian bread, and sprits," observing "we all lived upon ham and bacon."[33] Bacon was the most commonly consumed meat on the Oregon and California Trails. Immigrant Helen Carpenter complained about the monotony of overland trail food: "About the only change we have from bread and bacon is to bacon and bread." Authors of guidebooks for overland immigrants advised packing twenty-five to seventy-five pounds of bacon per person for the 110-day trek, which meant as much as over half a pound per day.[34]

Pork fueled the gold rushes, the logging frontier, military posts, and the canal and railroad boom across the continent. Packed pork, often of the lowest quality, was the principal protein for the forty-niners and other migrants. At the Hoff Store site in San Francisco, pig bones, cut and sawed at a New York packinghouse, dominate the archaeological remains from the Gold Rush period. Head and neck bones were present in large numbers, confirming that the miners consumed the less valuable parts of the carcass, including spareribs, shanks, and bacon. Miners complained about the quality of the pork they ate as well as the frequency. "The pork I bought in town last night is the stinkenest salt junk ever brought around the Horn," one miner noted. "It is a hardship that we can't get better hog meat, as it's more than half our living. We fry it for breakfast and supper, boil it with our beans, and sop our bread in the grease." Miners also ate beef (especially if they were lucky enough to strike the elusive mother lode), but salted and preserved pork was cheaper and readily available at remote mining camps.[35]

On the cotton frontier of the Southwest, people depended on pork more than did any other group of Americans. Much of the nation's wealth in this region as well as the nation was created by the labor of enslaved Americans, a group that grew rapidly in numbers and economic significance from the late eighteenth century to the mid-nineteenth. Slaves raised much of the nation's leading export—cotton—as well as tobacco, hemp, and grain. They constructed much of the South's infrastructure and industrial output. To the extent that enslaved hands built much of the South, much of the South was made by pork.

Slavery was so profitable for so many people for such a long time that planters recorded and discussed important aspects of slave management, including discipline, clothing, housing, health, and diet. Unsurprisingly, much of the discourse focused on food, specifically meat. Planters agreed that between three and a half and four pounds of meat per week, most often salt pork or bacon, was optimal for slaves. The amount actually served ranged from two to five pounds, but three and a half pounds was the most common ration. The escaped slave James Pennington recalled that the weekly provisioning routine on his Maryland plantation was issued by the overseer: "to every man the amount of three-and-a-half pounds to last him till the ensuring Monday." Englishman Richard Russell traveled to the United States and found that each able-bodied slave received half a pound of bacon per day as the meat ration. Frederick Law Olmsted reported on the 1850s slave diet on a large Mississippi plantation: "The allowance of food was a peck of corn and four pounds of pork per week, each. When they could not get 'greens' (any vegetables) he [the planter] generally gave them five pounds of pork." There were qualifications to the pork diet. "When we recommend bacon," a Georgia planter explained in 1857, "let it not be understood that salt provisions alone should be adhered to . . . an occasional mingling of fresh beef, or mutton or kid" was useful for proper digestion and health. Some slaves enhanced their diet with game and fish, and planters occasionally distributed other meat. Still, bacon and salt pork were the ubiquitous protein choices planters made about the slave diet.[36]

Most of the pork that enslaved people consumed was either fried or stewed. One-pot meals were common for the field hands who composed the majority of slaves on the large plantations, easing food preparation duties in the quarters. The slaves in the low country of Georgia and South Carolina prepared a dish called pileau (pronounced "perlu") made from the hog backbone boiled with rice. At butchering time, the organ meat and small intestines, known as chitterlings or chitlins, were cleaned and often fried.[37]

The evidence for meat consumption among free African Americans is more elusive, but what exists suggests that their diet closely resembled that of enslaved people. Bones found at the homes of free African American families in Alexandria, Virginia, indicate that pork (heads, stew meat, and feet) was the most common meat. Native Americans who consumed pork also tended to live low on the hog.[38]

White southerners, poor and wealthy, relied more on pork than on beef during the early years of the republic. In 1796, Latrobe described the condition of tenants and yeomen who lived in northern Virginia. They inhabited "miserable" log houses and kept their livestock consisting of a few pigs, fowl, and cow "at scarcely any expense in the woods." They consumed corn, bacon, milk, and vegetables in season, or, as Latrobe stated, "cabbages to their bacon."[39] Olmsted's travel accounts of the 1850s provide a good look at southern foodways among a cross section of the southern white population. From Virginia to Texas, pork in multiple preparations was usually on the tables at taverns and private residences. On a prosperous Virginia plantation, Olmsted dined on baked sweet potato, "four preparations of swine's flesh," fried fowl, fried eggs, cold roast turkey, and opossum. A supper in South Carolina consisted of "seven preparations of swine's flesh, two of maize, wheat cakes, broiled quails, and cold roast turkey." In Louisiana's Red River country Olmsted's hosts served cold, salt, fat pork, biscuit and corn bread, butter, and molasses. Meals in Eastern Texas included wheat bread with ham on one occasion and bacon, fried eggs, sweet potatoes, and bread on another. Beef was conspicuously absent from Olmsted's writings.[40]

Other travelers confirmed Olmsted's view of a pork-rich southern diet. Yankee John Hamilton Cornish reported from South Carolina in 1843 that "a dinner of hog, hominy, rice and sweet potatoes" was "the standing fare of the South."[41] Englishman James Stuart asserted that white southerners ate more pork than beef. A variety of meats was the norm for nineteenth-century southerners with aristocratic pretensions, as indicated by Stuart's dinner in Virginia with roast turkey, ham, roast beef, duck, game pie, and numerous side dishes. Most of the other meals Stuart consumed with white southerners, however, did not include beef. Stuart encountered ham and eggs as standard breakfast fare. Englishwoman Harriet Martineau noted that midday meals at prosperous residences consisted of roast turkey and ham, occasionally boiled fowl, and "a small piece of nondescript meat, which generally turns out to be pork disguised."[42]

Pork barbeque remained a distinct cultural feature of the South in the nineteenth century. Northerner Sarah Frances Hicks moved to North Carolina in 1853 and was impressed with the southern dependence on pork, especially barbeque. She noted that "red pepper is much used to flavor meat with the famous 'barbeque' of the South & which I believe they esteem above all dishes[,] is roasted pig dressed with red pepper & vinegar."[43] If anything, barbeque became more important to southerners as would-be aristocrats attempted to protect their positions of political, economic, and social privilege from dissatisfied and frustrated yeomen. Barbeques reinforced patterns of dependence, kinship, and patronage, serving as a reminder of shared cultural identity and white supremacy.

In the backwoods of the North and West, pork was also the primary meat choice. In western Pennsylvania during the 1780s, the German traveler Johann David Schoepf described travelers' meals: "The entertainment in woods-hotels of this stamp, in lonesome and remote spots throughout America, consists generally of bacon, ham and eggs, fresh or dried venison, coffee, tea, butter, milk, cheese, rum, corn-whiskey or brandy, and cyder." Schoepf also reported on a stop for refreshment near Pittsburg, where the host served bacon, whiskey, and cakes.[44] Francois Michaux traveled through Pennsylvania and Ohio in 1802 and commented on the frequency of ham served at taverns for breakfast and supper: "slices of ham fried in the stove, to which they sometimes add eggs and a broiled chicken."[45] A German settler in Indiana wrote home in 1836 that he and his party successfully butchered twenty-four hogs but very quickly tired of the monotony of eating pork.[46] According to William Cooper Howells, Ohio travelers of the 1820s encountered food that was similar to that of the South, "consisting of ham, eggs, chickens, turkeys, game, and now and then beef." At the fall cornhusking feasts of Howell's youth revelers enjoyed poultry and pork, not beef.[47]

Pork remained frontier food in the mid-1800s. Theodore Bost, a settler near Chanhassen, Minnesota, wrote to his parents in the spring of 1856 that "pork and potatoes and soup made from odds and ends gets terribly dull after a while."[48] James Creighton celebrated the Christmas of 1856 in Omaha, Nebraska Territory, with a meal of turkey "taken from the side of a hog."[49] A significant number of Missouri settlers relied on pork, since so many of them came from the Upper South where pork composed a major portion of the meat diet. Two excavated farm sites from Missouri's Salt River Valley show that pork constituted 95 percent of the domestic meat supply on one farm and 79 percent on the other.[50]

On northeastern farms, pork continued to play a significant role, albeit a secondary one. Asa Sheldon of rural Massachusetts recalled that in his youth at the turn of the nineteenth century brown bread made from cornmeal and rye flour was the everyday staple year-round. In the winter, beef broth and porridge were common. The porridge was made by boiling a piece of pork, adding beans, and cooking to a paste. "Our Sunday dinner," the elderly Sheldon recalled in 1862, "was invariably baked beans with salt pork, and a baked Indian pudding."[51] The iconic New England dish of baked beans is a good example of how the diversity of northeastern mixed farms mitigated against pork as the most important meat. Beans were a plentiful garden and field crop, unlike in the South and West, where they were primarily a garden crop. Molasses was readily at hand in New England since that region had long been engaged in distilling molasses from the West Indies into rum. The salt pork in the bean pot, however, provided more flavor than protein and calories.

In northern cities, beef and mutton were important because consumers could purchase fresh cuts and consume them before they spoiled. Fried steak was a frequently prepared dish at home and in restaurants.[52] During the Christmas season of 1864, *Harper's Weekly* reported that New York butchers had received "nearly 7,000 beeves, about 27,000 sheep, and as many swine" for the holiday season. The dressed weight of those 7,000 cattle likely exceeded the total weight of 27,000 dressed hogs by a significant margin.[53] While beefsteak was most important on the bill of fare in urban homes and restaurants, pork was still common. A Russian traveler noted that in "New York as well as in Cincinnati not a single meal in a hotel is without this meat [pork] in various forms and preparations."[54]

The preference for beef among the diet of the rapidly growing urban population in the North affirmed the English preferences and reflected the growing class sensibility. There had always been a "middling sort" of people in America, but the segmentation and specialization of the economy symbolized by the factory meant more jobs for managers and professionals such as lawyers, increasingly known as the middle class. Members of this class preserved their social distance from the working masses through diet. As Eliza Leslie noted in her 1857 cookbook, pork spareribs were seldom served at "good" tables and the practice of eating meat from the head or gravy made from pig brains was an "unfeminine fancy."[55]

Pork defenders fought back. As one writer noted in 1859, "It is very common to hear flippeant [sic] girls, would-be delicate women, and would-be very nice young men, who live in cities . . . declaim against pork." Pork, the author stated, was the meat that built the nation. "Nearly all the gallant men who have served our country in her times of greatest peril were raised on pork," he argued, listing numerous notable leaders of both the French and Indian War and the American Revolution. Grain-fed western pork was pure food, in contrast to city-consumed pork made from offal- and garbage-fed hogs. Quality pork supplied the "bone, muscle, and sinew, the brain and nerve which is to be the strength and virtue of our future history."[56]

## WARTIME PORK: SOWBELLY, SALT, AND SCARCITY

Just as pork sustained Revolutionary War armies, Civil War soldiers regularly ate pork as part of their government-issued rations and recovered it from smokehouses and poached it on the hoof. The US commissary issued beef and pork, but salt pork, commonly known as sowbelly, was the most frequently issued meat in the US Army. Much of this meat was prepared by frying. Skillygally was a ubiquitous dish, consisting of hardtack fried in the fat of the salt pork. Warren Freeman described the meal in a letter to his father in early 1862. Freeman recalled good home-cooked meals but "freely put them by, and fry my slice of salt pork, which with a bit of ship-bread, satisfies my necessary wants."[57]

The Confederate commissary struggled to provide enough pork and salt, which meant that southern soldiers, having developed a dependency on pork, ate more beef during wartime. The loyalty of Kentucky in 1861 and loss of central Tennessee in 1862 denied the Confederacy access to the South's major hog-producing section. Furthermore, some southern planters resisted raising food crops and livestock in favor of cotton. The lure of high cotton prices and the expectation of a short war likely militated the transition to food production. There was also the issue of planter identity. One Georgia newspaper editor decried the haughtiness of planters who would "rather raise a pound of cotton at three cents than a pound of bacon at a dollar."[58]

A hog cholera epidemic further hobbled Confederate self-sufficiency. Mississippi reported significant losses in 1862, Alabama and Georgia in 1863, and Virginia, Florida, and Alabama in 1864. Alabama planter James Mallory noted in his diary on September 15, 1863, that due to hog cholera "bacon will be scarcer than ever known." On September 27, Mallory reported, "The disease among the hogs continues, it has made its appearance in our stock and will likely destroy most of them, it prevails throughout the Confederacy, the meat question is settled, the people will have to do without it, the army must be supplied."[59] Conditions were so poor that some planters protested that they were unable to meet their own food needs, let alone provide the tax in-kind that Congress required in 1863.[60] While the losses due to hog cholera were real, it is possible that southerners may have exaggerated their losses to protect their assets from taxation and impressment, although men such as Mallory had little incentive to dissemble in personal diaries.[61]

The South also experienced a salt shortage that complicated the war effort. Salt was necessary to preserve meat, but prices doubled by late 1861 and remained high for the duration of the war because the Confederacy was cut off from domestic supplies in western Virginia and from imported salt. David Anderson of Spartanburg County, South Carolina, wrote to his brother in Charleston to purchase salt for five dollars a bushel in mid-1862. Given the shortage, Anderson wanted to have some on hand "to be certain of it for Killing Hogs" in the fall. He claimed that it was preferable to purchase the salt in Charleston than to travel to Virginia to obtain it.[62] A Mississippi planter reported in late 1862 that "we have corn for bread; we have hogs for meat, but no salt to save it."[63] Historian Joe Mobley noted that by 1863 the salt crisis resulted in a significant shortage of preserved pork in the forms of bacon and ham across the Confederacy, exacerbating the meat protein problem.[64] As one North Carolinian wrote to Governor Cyrus Vance, "Blessed are they that have no hogs," for they would not have to pay the outrageous sums for salt to make bacon and pickled pork.[65] As a result, many southerners ate more fresh than preserved pork during the war years.

Southerners attempted to manufacture their own salt, with mixed success. In October 1862, William J. Grayson noted from Charleston, South Carolina, that "the world is agog for making salt," with numerous kettles in operation along Charleston's waterfront. According to many critics, Grayson observed, locally made salt was not suitable for curing meat, but he related a report from a Mr. Collman that "his bacon last winter with salt of his own making and the hams he is now eating from it are as good as any he has ever tasted." The price of salt in the city suggested that the local salt boom was profitable, at least for some Charlestonians, even if it failed to ease the shortage.[66]

Changes in US military policy compounded the southern pork problem. As the war dragged into a second and then a third year, President Lincoln and military leaders proved increasingly willing to carry the war to the civilian population. Foraging parties decimated the hog and pork supply wherever they went. Federal soldiers celebrated their proficiency in killing and butchering hogs in their accounts of wartime service. One general observed, only half-jokingly, that the men of the 14th Illinois "could kill, skin, and fry a hog and put it in their haversacks without breaking ranks."[67] When a hog passed the guard of a US Army camp in 1864, a

**Figure 17**

Salt pork was a critical component of US Army rations during the Civil War, but soldiers also supplemented their diet with fresh pork, as seen here in Alfred Waud's sketch *The Bivouac Feast*, depicting US soldiers roasting a hog. The presence of open-range swine and the destruction of fences by advancing armies meant that there was fresh pork available for both the Confederate and Union armies. Source: Library of Congress. Public domain.

Pennsylvania captain managed to jump on its back and slit its throat. "Without waiting to dress it," the memoirist related, "he began cutting off pieces and throwing them to the crowd. The smell of fried pork soon pervaded the camp," and within fifteen minutes the meat was gone.[68]

By late 1864, General William T. Sherman, promising to "make Georgia howl," hoped to chasten southerners for the error of secession. In October 1864, Sherman asserted, "When the rich planters of the Oconee and Savannah see their farms and cow and hogs and sheep vanish before their eyes they will have something more than a mean opinion of the 'Yanks.'"[69] Sherman's army subsequently cleaned out all kinds of livestock and meat in the March to the Sea and the campaign through the Carolinas, but hogs were among the most commonly found, killed, and enjoyed livestock by Sherman's bummers in large part because the campaign coincided with the penning and fattening season for hogs. On November 16, just two days into the March to the Sea, Iowan John Rath foraged near McDonough and brought in "one fat porker." Two days later, he observed "country good, plenty of forage such as sweet potatoes and pork." When the army crossed into the more sparsely populated coastal plain southeast of Macon and Milledgeville, however, they found less food. That December, Rath reported crossing expansive tracts of forest and swamp with few plantations and made no references to meat. On December 20,

Rath declared that the men were "almost starved," a stark contrast to the rich living of November.[70] Near Savannah, however, foraging improved as the men encountered the cotton and rice plantations of the coast. An Indiana soldier noted that after one late December expedition he ate so much ham and sweet potatoes "I thought I was hollow clear to my toes."[71]

Confederate troops contributed to the southern pork famine. They had little hesitation about taking from fellow southerners, leading some civilians to complain that the Confederate Army was as unwelcome as a Yankee. In September 1864, General Lawrence Ross's Texans made themselves notorious near Fayetteville, Georgia. "Citizens complain almost daily of the depredations of this brigade upon their property," wrote the assistant adjutant general. "Hogs are being stolen and killed almost every night, and it is certain that some of the officers of the brigade are acquainted with the names of the offenders." One former slave from northern Alabama recalled that Confederate troops took the meat and meal from the plantation and US soldiers subsequently took the hogs.[72]

Enslaved people were partially responsible for the depletion of southern herds. For generations planters feared that slaves poached livestock, but the extent to which their fears were grounded in the reality of theft versus the fear of a hostile, internal enemy was debatable. Regardless of the reality of such concerns, the wartime chaos made it easier for enslaved people to supplement their diets with free-range animals. In early 1863 a group of African Americans engaged in a hog hunt, described by US General John Dix as "catching hogs running at large in the woods and carrying them off to kill them."[73] In upcountry South Carolina, one woman reported a rumor that the slaves in Spartanburg were "killing up hogs and anything they can get hold of. Mr. Judd has had to lock up his hogs."[74] It is unknown if enslaved people near Spartanburg really stole as many hogs as this woman claimed, but the fear of theft shaped white southerners' attempts to secure their food supply.

Like Mr. Judd of South Carolina, many southern farmers attempted to hide their animals and meat from enslaved people, deserters, loyal citizens, and invaders. Georgians sometimes suspended ham and bacon from tree limbs away from easily pilfered cellars, storehouses, and smokehouses.[75] When US soldiers in Virginia passed Alberta Gallatin Sale's Northbank plantation, Sale complained that "they took all my bacon except what I hid."[76] A South Carolina soldier wrote from near Gordonsville, Virginia, in April 1862, "The farmers of the country behind us are driving their hogs and sheep, droves of them, toward Richmond, to prevent them from falling in to the hands of the Yankees."[77] Hogs hid themselves as best they could, but hiding was difficult when the enemy was nearby. A Hoosier officer of the 14th US Colored Troops approached a plantation house near Chattanooga and asked "a very fine looking lady" if she had any livestock. She replied that General Rousseau's men had passed the day before and took everything. The officer was ready to depart when one of his men reported that there were hogs in the orchard, which the men promptly rounded up.[78]

The Civil War brought the nation's historical pattern of dietary preferences into relief. English and European colonists successfully transplanted their meat hierarchy in America. Across the colonies, beef was the most sought-after meat, followed by mutton and then pork. By the mid-nineteenth century, however, pork had assumed special importance in rural areas, most notably in the states that made up the South and West and the borderlands. The wartime experiences of many US soldiers revealed the decline in mutton and rise of pork in these regions. Missouri soldier Albert Demuth, himself a westerner, found that people in rural Arkansas were especially partial to pork. He reported to the folks back home that local "Secesh girls" "never eat anything but cornbread and hog meat."[79] Indeed, people on the settlement frontier in Iowa, Kansas, Oregon, and countless other places relied on pork. It was a mainstay for landless tenant farmers, enslaved people, and property-owning yeomen as well as the urban poor, lumberjacks, soldiers, and sailors. One of the continuities of English and European migration was that beef and mutton were the meats of choice, although mutton lost its place in America. For American laborers, like the peasants and yeomen of the Old World, hog meat was an ideal food.

CHAPTER 4

# PIGS AND THE URBAN SLOP BUCKET

In 1788, a satirical petition to New York City's leaders from "the SWINE belonging to said city" appeared in *American Magazine.* Hogs of the city, according to this "pig" with a pen, were "natural born swine and peaceable subjects," that suffered "numberless abuses and violent unmerited outrages" at the hands of the city's human inhabitants despite their "quiet behavior and faithfulness . . . as scavengers of the streets." Coachmen ran over pigs, porters and draymen whipped them, and kitchen maids poured boiling water in the streets, scalding snouts. Furthermore, "waggish boys" rode hogs for sport. All of these outrageous actions interfered with the service work pigs performed as scavengers on behalf of the city and its inhabitants.[1] They deserved better.

Swine in the city were not unique to New York. For hundreds of years, they were part of urban landscapes in Europe, cleaning the streets of kitchen waste, insects, carrion, and manufacturing waste such as brewer's mash and slaughterhouse offal. While city pigs created their own waste, it was not especially offensive or difficult to manage compared to the problems they solved. After all, horses, dogs, fowl, cattle, and humans—all animals in the city—deposited their manure in the streets, alleys, and vaults.

Town and city leaders in the American colonies created laws to control the movement of animals, but they enforced them with varying degrees of vigor and experienced irregular success. Hogs successfully exploited a niche in preindustrial cities and, with the rise of industrial capitalism and emerging class sensibilities, found themselves forced into narrower confines on the edges of cites. Despite campaigns to purge hogs from urban settings, they remained a critical part of the city until recently. Even today, hogs help city leaders transcend environmental limits of waste disposal, just as they did in 1788 when New York's hogs aired their complaints.

Another satirical letter to the editor indicates the ubiquity of urban pigs in America. In 1809, a letter from "PIGGY" to the editor of a Philadelphia comic paper suggests that the hogs were not altogether benign. The author praised the fine embroidery produced by Philadelphia's women, noting that a young woman emerged from her house with a "neat dress, worked with an oaken wreath replete with acorns" but walked no more than a block from her home before returning, "owing to the number of swine that were attracted by the sight of the acorns." The writer blended images of traditional rural mast-feeding practices with the scavenging diet of urban pigs to demonstrate the pervasive, if inconvenient and sometimes annoying, presence of hogs in the city.[2]

Most people who lived in early America understood what historian Joel Tarr has labeled the metabolism of the city, even if they would not have employed the same vocabulary. Tarr explained that the flow of resources between city and hinterland and the processes of converting those into something else resembled the metabolism of a body. The author of an editorial in Philadelphia's *American Museum* in 1788 preempted Tarr. Upon learning of a new law that prohibited free-range pigs in that city, the writer explained, "Nature does nothing in vain. She is a great economist in all her works. She appears to have intended hogs to feed on those offal matters which would otherwise become not only offensive to two of the senses, but the cause of putrid diseases." The prohibition, according to the editorial, increased waste problems "more in the city than ever; and hence arises the usefulness of hogs in our streets. They kindly supply . . . the want of a city government."³ Urban leaders alone were incapable of ridding the streets of organic waste that was the source of offensive smells and disease. In New England's coastal towns, where fish was a significant part of the urban diet and economy, hogs consumed the waste from commercial and household fish processing. In these places, forward-thinking citizens labeled hogs as "physicians" because they prevented the spread of disease by removing fish waste. Operators of urban distilleries and breweries kept swine to consume the distilled grains and mash that would otherwise have constituted a major urban disposal problem.

Hogs also provided meat that fed the poor of American cities. The Philadelphia editorialist claimed that the custom of free-range pigs permitted "a number of poor people to lay up a few pounds of salt meat for the winter."⁴ The author estimated that with each animal gaining approximately fifty pounds over the summer, over forty thousand pounds of meat were harvested from the streets to feed Philadelphians who would otherwise pay more for meat, consume less of it, or go without. New Yorkers who opposed efforts to rid the city of hogs insisted that the income from the sale of the family pig or the meat it provided was a hedge against hard times, even preventing families from becoming a "public burden" by relying on local welfare.⁵

## TAKE CARE OF THE PIGS: CONFRONTING URBAN SWINE

Visitors to the United States understood the waste disposal and food supply aspects of hogs in American cities, even if they were not always comfortable with animals in the streets. In 1820, Englishman Charles Henry Wilson commented that visitors to New York City were "continually annoyed by innumerable hungry pigs of all sizes and complexions, great and small beasts prowling in grunting ferocity, and in themselves so great a nuisance, that would arouse the indignation of any but Americans." Norwegian Ole Munch Ræder reported from the United States in 1847, "I have not yet found any city, county, or town where I have not seen these loveable animals wandering about peacefully in huge herds." According to Ræder, hogs cleaned the streets by "eating up all kinds of refuse. And then, when these walking sewers are properly filled up they are butchered and provide a real treat for the dinner table." Charles Dickens toured New

York City in 1842 and simply cautioned readers of his *American Notes* to "take care of the pigs" that were hard at work as scavengers.[6]

Urban swine also irked Americans. An 1806 complaint to New York City's Board of Health noted that two pigs kept in a shoemaker's six- by three-foot backyard sty were the source of a nauseating summer smell. It was so bad that the complainant had to close his windows, which was equally intolerable due to summer heat and lack of air circulation. In this case, the real issue was not pig keeping, but that the offending neighbor violated norms by permitting the pigs to deposit manure at home without removing it, keeping the stench at home too.[7]

It was not just odor that generated complaints. In 1825, a hog in Stonington, Connecticut, ventured into a storage cellar near the waterfront and pulled the wooden taps from two hogsheads of molasses, permitting the liquid to flow across the floor. Residents chased the hog out, but the editor of a city newspaper used the occasion to call for "enforcing the by-law for restraining swine from running across the borough." A newspaper editor labeled New York City as "One Grand Piggery" in 1830. "It is not unfrequently the case," he noted, "that strangers while turning the corners of streets have come into contact with, or tumbled over these unmannerly brutes, and risen either with bloody noses, or with clothes covered in filth." The editor closed with a complaint that "the government of the city justly deserves censure for permitting these things to be." Increasingly, critics called for the enforcement of existing laws or the enactment of new ones to control city swine.[8]

Numerous city governments attempted to exercise control over what many people viewed as a growing problem. An ordinance to prohibit swine from running at large in Chillicothe, Ohio, was strengthened in 1802, allowing anyone to collect stray pigs and present them to the city marshal who would impound the animals, sell them at auction, and then divide the proceeds equally between the marshal and the person who captured the stray. In 1809, the city council in Washington, DC, required ringing free-ranging hogs and imposed fines on owners of unringed, stray hogs. The ordinance authorized citizens to take and kill unringed hogs, with the claimant reserving half the meat and donating the other half to the city for poor relief.[9]

Evidence from Philadelphia and New York City suggests that these ranging hogs actually belonged to poor, unskilled laborers or members

## PIGS AND THE URBAN SLOP BUCKET

of the growing class of mechanics such as shoemakers, bakers, and sail makers. The latter earned better wages than unskilled workers, but remained vulnerable during the inevitable economic downturns that threatened them with accepting public relief or moving to the poor house. As a hedge against hard times, many families kept a pig in a small sty, allowing it to scavenge on waste on the streets away from home for much of the summer months. This was ideal for the working poor and their sympathizers, but it came under increasing fire from reform-minded citizens in the nineteenth century.

New Yorkers escalated the fight over pigs in 1816. That year, citizens presented petitions to the Common Council to forbid individuals from allowing hogs to run free, ostensibly to prevent damage to the streets. After the petitions' defeat in June 1817, the council voted again and passed an ordinance that autumn, initiating a torrent of appeals and protests. When Mayor Cadwallader Colden, also a city judge, impaneled a grand jury on the issue, two citizens were charged with the misdemeanor of "keeping and permitting to run hogs at large" in the city limits.[10]

The 1817 trial of Christian Harriet in New York City provides insight into the contest over pigs in the streets. Harriet was a skilled Manhattan butcher who was one of those two men who owned hogs that ranged freely. Prosecutors claimed that the hogs were a nuisance because they attacked children, defecated on people, and copulated in view of women. Counsel for Harriet countered that keeping pigs was a custom of "immemorial duration" that constituted neither a misdemeanor nor a nuisance. The defense claimed that the state, not the city, possessed jurisdiction regarding pig keeping. Judge Colden instructed the jury that removing swine was necessary to achieve a healthy city. For those concerned about the working poor, Colden believed that good jobs and wages were the best safeguard against hard times, not pork on the hoof in the streets. Colden envisioned a place where women could "walk abroad through the streets of the city without encountering the most disgusting spectacles of those animals indulging the propensities of nature." It was a vision of propriety that reflected the class sensibilities of upwardly mobile and aspiring New Yorkers, a vision the jury shared. The court found Harriet guilty and fined him one dollar and court costs. The token fine was likely because he was guilty only of permitting his hogs to run free and that there was no specific damage claim associated with his animals.[11]

In the years after Harriet's trial, New Yorkers resisted the numerous attempts city leaders made to clear the streets of swine. Protestors, mostly Irish and African American, greeted a city-ordered roundup in 1821 with a mob and freed the impounded pigs. The scene was repeated in 1825, 1826, 1830, and 1832. When cholera swept the city in 1832, public health officials targeted pigs and garbage because the disease hit poor and working-class districts so hard. Health concerns, not concerns for propriety or property damage, justified hog removal. To pig owners, however, the new justification was meaningless; they valued their waste-fed pork. Police rounded up pigs in the face of continued resistance from owners and, presumably, the pigs. By June 1849, police successfully gathered approximately six thousand swine and drove many thousands more northward beyond the city limits. Even then, however, pigs remained at large on the streets.[12]

An 1849 cholera outbreak strengthened the reformers' position. The supposed link between hogs and cholera was explicit for readers in newspaper articles such as "The Cholera and the Hogs." A writer for the *Brooklyn Daily Eagle* claimed that hogs, "alive and squealing," would soon be visited by their "friend, the Cholera."[13] To control the spread of cholera as well as to clean up the city, regular pig roundups took place throughout the 1850s, with varying degrees of success. The *Daily Eagle* labeled the 1849 roundup campaign as the "Pig War." A group of city marshals "assisted by some dozen negroes" ventured into an Irish settlement near Warren Street. They started tearing down pigpens and were subsequently attacked by a group of men, women, and children wielding clubs, sticks, and stones. The pigs complained, too, in the form of a letter to the mayor from "Little Spot" published later that year. Little Spot wrote from the pound, lamenting the fact that he was fed on slops. He missed "dainty bits from the rich man's table" such as a "nice piece of roast beef, and now and then a pickle" and pledged to behave, if released.[14] Little Spot's protest notwithstanding, the war continued throughout the 1850s. In 1854, a writer for the *New York Times* claimed that Mayor Lambert and the people of Brooklyn were "down on hogs," but resistance continued when owners of pigs, goats, and cattle attacked police officers on roundup duty in 1856.[15]

There was a difference between the clean-up campaigns of the 1850s and those of the 1820s. The target of the new effort was more likely to be a piggery than simply free-ranging animals. Piggeries were complexes of enclosures where the hogs consumed garbage and by-products of slaughterhouses, distilleries, and dairies. New York City's pig war intensified in the summer of 1859. The city's Public Health Commissioners met in July 1859 and heard the chief health inspector's cleanup plan. The city provided notice to piggery owners, as well as those who boiled offal and butchering waste, and residents between Fiftieth and Fifty-Ninth streets and Sixth Avenue and Broadway to move beyond the city limits. Less than a month later, police conducted a raid in the twenty-second ward (north of Fortieth Street and east of Sixth Avenue). Eighty-seven men including police officers, health wardens, meat inspectors, dock watchmen, and others headed for the piggeries, "armed with pistols, clubs or daggers, and a few carried pick axes and crow bars with which to demolish the pig pens." They approached numerous owners (several of whom had already removed their pigs and whose enclosures were subsequently torn down) and spread lime to reduce odor. Several owners had not moved their animals. Those hogs were "carried into the street" and impounded.[16]

Reform-minded New Yorkers claimed a partial victory in the 1859 "Piggery War." They impounded approximately nine thousand hogs, although owners subsequently recovered most of them. Some owners relocated to Westchester County and Brooklyn, causing an increase in complaints in those areas. After several more sweeps in the following weeks, the inspector managed to shut down the remaining piggeries, having destroyed three thousand hog pens and spread two hundred fifty barrels of lime, no doubt to the relief of residents on the west side of Central Park. To secure a lasting victory, the city inspector proposed a fine of twenty-five dollars for every pig kept south of Eighty-Sixth Street. Those who kept pigs north of this point without a permit were subject to fines.[17]

Stories of urban pigs in American history invariably featured New York City, but why were the city's pigs so notorious? The pig problem here looked so enormous simply because the city was so big, over three times larger than the nation's second largest city, Baltimore, in 1850. The population was growing rapidly during the 1840s and 1850s, with thousands of fresh arrivals from the German states and Ireland fleeing economic and political marginalization. With so many people, especially the poor, in search of economic security, it is understandable that there were a great many hogs. A New York observer estimated that there was one hog for

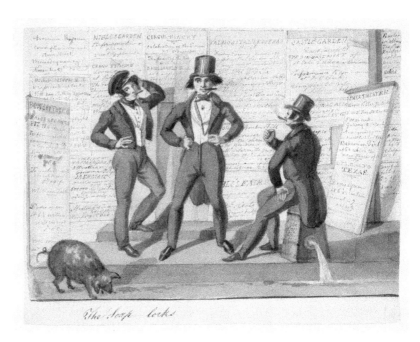

**Figure 18**

Pigs were omnipresent in American cities during the nineteenth century. This watercolor, titled *The Soap Locks, or Bowery Boys* (ca. 1840), by Nicolino Vicomte Calyo, linked the emerging political culture of ethnic groups and youthful gangs with urban disorder, represented by the hog in the gutter. Pigs were vitally important for working-class people as a hedge against the frequent and often catastrophic economic downturns of the early to mid-nineteenth century. Middle-class city leaders disapproved of free-ranging hogs as a threat to their vision of a clean, sober, and decorous city. Source: Yale University Art Gallery. Public domain.

every five people in 1820, suggesting a population of twenty-five thousand animals. If that ratio held in 1850, the city would have been home to over a hundred thousand hogs. An area near Manhattan's present-day 125th Street acquired the name Pig's Alley due to the large number of hogs there. The new immigrants lived on the city's physical margins and relied on urban animals. As we have seen, the hog furor was also partly a result of the growth of the city's middle class. The new, reformed cities were to be places "of parks and promenades," as historian Catherine McNeur stated, safe places for middle- and upper-class residents to engage in leisurely display. Political leaders needed to placate this group to govern the rapidly growing and unruly metropolis.[18]

The frequent emphasis on New York City accounts in the story of America's urban pigs obscures the fact that they were common fixtures in urban streets, alleys, and house lots across the country. In 1847, the city of Baltimore allowed any person, "white or black," to take stray pigs and use them without penalty. Despite an 1836 law that prohibited swine from ranging at large in Washington, the problem persisted in 1850, when a wag parodied the sectional crisis of that summer and wrote that the city needed a "compromise line" to create free territory for swine to run at

## PIGS AND THE URBAN SLOP BUCKET

large. In Bangor, Maine, a writer stated in 1852 that swine constituted an "intolerable nuisance." San Francisco residents petitioned their city government in 1858 to enforce the statute prohibiting swine from running at large. They noted that the city's public places were "still infested by swine and goats, against whose depredations no fence nor any amount of care is any security" and requested the employment of a hog and goat catcher to gather and impound strays.[19]

In Pittsburgh, the city's first law to prohibit swine running at large was enacted in 1794, but aside from an amendment in 1816 to fine those who let animals loose in the city limits, there was little agitation for reform until 1849. That year, the city pledged that municipal officials would enforce the hog law. Two years later the city offered a dollar bounty per animal. Financial inducements aside, it was the subsequent hog drives that helped reduce the threat of obnoxious pigs. The city's 1859 roundup, according to historian John Duffy, was "the most effective."[20]

Boston's mid-nineteenth-century experience was somewhat more orderly than that of New York and Pittsburgh. The Boston Board of Health (constituted in 1799) determined that a license was required to maintain a pigsty. Licensed keepers could have pigs in the city between May 1 and October 1 if they placed their sty over moving water to carry away waste.[21] As early as the 1830s, the city's piggeries were beyond the city limits, militating against complaints of pollution and smell. A piggery located in West Cambridge, six miles from Boston, received waste carted from Boston residences. The hogs existed "entirely on the offal from the dwelling houses of Boston." The piggery owners maintained over seven hundred hogs on fifteen acres and paid the city $2.75 per load of waste.[22]

The pig purges from the major cities pressured surrounding areas. After Brooklyn's pig war, Flatbush experienced the growing pains. A belt along the northern border of Flatbush three miles long and one-third of a mile wide was the site of extensive garbage feeding operations for Brooklyn's waste. Dr. H. Bartlett, the local public health officer, inspected those piggeries in 1874 and described the "piles of manure, mixed with contents of privy vaults, distilled hops, decayed meats, and vegetables" that generated powerful odors.[23]

The debate about the Flatbush piggeries continued, highlighting concerns about the appropriate place for hogs as well as class and ethnic conflict. Local elites, predominantly descendants of Dutch immigrants, singled out the Irish for perpetuating the piggeries, although the reality was more complex because the Dutch owned most of the land, including that occupied by piggeries. In 1882, the Flatbush Board of Health declared the piggeries a nuisance and ordered that they be shut down no later than January 1, 1882, along with a prohibition against keeping more than two hogs on any property without permission of the board. The piggery keepers, however, evaded capture by driving their herds back and forth across the line dividing Flatbush and Brooklyn. Brooklyn officials found that the hogs moved to Flatbush, and Flatbush officials found that the offending pigs relocated to Brooklyn. In 1885, a coordinated effort on the part of officials from both cities resulted in the capture of the piggery operators. Law enforcement officials, armed with clubs and revolvers, persuaded the piggery operators to go before a judge.[24]

By the late nineteenth century, towns around Boston struggled with hogs. Leaders in Needham, located on the city's outskirts, became uncomfortable with piggeries in the 1890s. One of the most notorious piggeries, located just a mile from the center of town, was home to hundreds of swine. "Located near the railroad," one observer noted, "the stench would waft into the windows of cars every time a train passed; the passengers would close the windows, cage the smell, and carry it with them all the way to Boston." Local residents also suffered, which split the town along the lines of a hog party and an anti-hog party, with the latter desirous of ridding the town of piggeries. While the anti-hog party opposed piggeries, it supported the rights of individuals to keep a few pigs in what it described as "a Cleanly Way." They carried the 1892 election for selectman and attempted to put the piggeries out of business by issuing public health orders to shut down the piggeries. The operators apparently did so, although one of them resumed operations on a small scale. When a reporter for the *Boston Daily Globe* approached the house of the piggery owner, he found only a hired man at home who claimed that the owner was in Dover, just southwest of town. Needham, like Flatbush, Brooklyn, and New York before, simply kicked the pig issue down the road.[25]

Swine, having ensconced themselves in urban landscapes, proved to be far more persistent than reformers imagined and at least as troublesome as they were persistent. An editor from Bangor, Maine, complained in 1866 about disgusting odors that emanated from the hog pens in his city, many of them "kept in the most filthy state." "Let one lie down at night," he observed, "with open widows and breathe the effluvia arising from a pen containing fifteen or twenty hogs within thirty feet of his sleeping room, and if he wakes in the morning refreshed, it must be because he has more of the characteristics of swine than most of us are willing to admit."[26] He urged the city to pass an ordinance to prohibit urban swine keeping. In Milwaukee, city pigs were legal, although by 1884 they constituted a nuisance in the extreme northern and southern parts of the city. As the city health commissioner stated, there was a significant risk of disease from the garbage piles that were the feed for these city pigs. He urged aldermen to prohibit the keeping of pigs within the city limits, less because of the pigs and more due to their food source.[27]

Even small towns debated the presence of pigs on the streets. In Blowing Rock, North Carolina, the rise of middle-class sensibilities among city leaders contributed to control efforts. Perhaps more importantly, the increasing importance of a tourist economy based on clean mountain air and scenery was an inducement to prohibit pigs from roaming at large in town.[28]

## GARBAGE FOR PIGS

During the late nineteenth century reformers pushed pigs to the margins, but they did not end the reliance on pigs for garbage disposal in American cities. Organic waste from restaurants and hotels, known as market refuse, ended up as hog feed, fed raw or sometimes cooked and reduced to grease and tankage. Solid items such as metal and glass were often dumped or salvaged. Rubbish, the remainder of city waste, was frequently incinerated, dumped, or buried. Given the expense of incineration and reduction facilities, however, burial and feeding were favorable low-cost disposal options for organic waste in America's booming cities that produced a booming waste problem.

Yet the sanitary and class sensibilities of the nineteenth century were even more entrenched in the twentieth. Historian Martin Melosi noted that while garbage feeding faded in importance after 1900, concerns over the food supply during both world wars boosted the practice. By the 1940s, many cities relied on hogs to reduce their waste.[29] Despite the long-standing complaints about piggeries, they remained part of the urban and suburban landscape. The difference was that they were located at increasingly greater distances from the city.

Southern California is a good example of how cities utilized pigs to solve their growing waste problems in the twentieth century. As early as 1901, the *Los Angeles Times* reported on the problems at Long Beach. "Time was," the correspondent noted, "when it was considered the proper thing to make a dumping place of tide flats west of town," but objections from neighboring Alamitos and the fact that the tides simply returned the trash to Long Beach made this solution untenable.[30] The people of Los Angeles then hauled garbage by rail to a two-thousand-head piggery located between El Monte and Covina in Los Angeles County. By 1910, however, there were mounting complaints about odor from Angelenos who lived near the garbage-loading station on the Covina line as well as residents of Baldwin Park and Covina. A delegation of city officials, including the mayor, health inspector, and council members, visited the dump and witnessed garbage fed to hogs. Their report, however, was not reassuring to those who complained of the stench. They concluded that the "location and internal conditions" were "without offense" and that both garbage and hogs were "as sweet as they could be."[31]

Conditions at Covina, however, were hardly sweet. In July 1911, a grand jury determined that the hog ranch, located on the banks of the San Gabriel River, was a menace to public health. Interurban passengers complained that they were "running the typhoid gauntlet" as they passed the Covina station. The potential for contamination of surface water and the putrid odor from the loading station and feeding operation were severe. In 1912, the inspector of the Board of Public Works recommended constructing an incinerator at San Pedro to handle up to 260 tons of garbage per day. Los Angeles constructed the incinerator in 1914, but continued to rely on hog farms for disposal of market refuse.[32]

During World War I, Americans turned to garbage feeding with renewed energy. In a fit for efficiency, garbage feeding promised to reduce waste and increase the food supply, solving two national problems in the name of progressive conservation. As a US Food Administration publication scolded, "The American garbage pail, with its twenty-odd billion pounds of garbage per year, can well be considered one of our expensive luxuries." Even a modest reduction would result in "no small additional supply of foodstuffs."[33] Wasting garbage in wartime, the author noted, was just as indefensible as wasting food. A US government survey indicated that leaders in a fifth of American cities with populations of ten thousand or more considered altering their waste disposal techniques. Hog producers took note. John Evvard, a hog expert at the Iowa Agricultural Experiment Station, explained that army camps that housed between thirty and fifty thousand men would be suitable candidates for garbage feeding.[34] As one writer observed in 1918, "Garbage as a feed for swine is being utilized and sought after at the present time to a greater extent than at any previous time" due to high pork and feed grain prices.[35]

The wartime garbage-feeding surge increased concerns about the spread of hog cholera. One expert explained, "Garbage contains the infection or causative agent of hog cholera at practically all time." Pork from hogs slaughtered during the cholera incubation period could end up in the garbage supply, only to infect more hogs. War Food Administration officials reminded feeders that hog cholera immunization "is absolutely required." This was especially true for Massachusetts, where garbage-fed herds kept near cities constituted 90 percent of all hogs in the state. After adopting a statewide immunization program, Massachusetts all but eliminated hog cholera losses.[36]

The demand for garbage feeding continued after the war, thanks to accelerated urban growth. In 1921, Los Angeles solicited bids for hog feeders and ultimately contracted with Fontana Farms of southwestern San Bernardino County. City officials and Fontana Farms agreed that Fontana would pay sixty cents per ton of garbage for approximately two hundred tons per day, commencing on September 1, 1921. Fontana promised to haul the garbage across the county in "closed, specially constructed steel railroad cars" at no risk to citizens.[37] Work was soon under way at Declez to accommodate fifteen thousand hogs. The new "ranch" would include a line of pens stretching for three-quarters of a mile on either side of a railroad siding.[38] A steam-powered shovel removed the waste from the rail cars, which were then pressure-washed before returning to the city. Fontana Farms provided supplemental feed, consisting mostly of ground barley. In 1926, Fontana Farms bred a thousand sows and sold over forty-two thousand market hogs. Those animals produced up to forty tons of manure per day, which ended up as fertilizer on Fontana Farms land.[39]

CHAPTER 4

**Figure 19**
Dealing with urban food waste was a major problem for cities and surrounding areas throughout the twentieth century. Efficiency was the key to success at San Bernardino's Fontana Farms from 1921 to 1951. Specially designed railroad cars filled with waste arrived at Fontana Farms, where a steam locomotive and shovel delivered the garbage to the feeding pens closest to the rail line, circa 1935. Source: Los Angeles Public Library Photo Collection. Used with permission.

The garbage-fed pork from Fontana and other locales had problems. Almost all observers noted that the quality of pork from garbage-fed operations was inferior to that of farms that utilized grain and forage rations. Garbage-fed hogs experienced greater shrinkage between the lot and packing plant, and the dressed carcass had a lower percentage of meat. The meat was also flabbier, softer, and oilier than grain-fed pork. When the pigs ate decayed or fermented garbage, the pork was likely to be "strong and rancid," ending up in sausage to disguise the flavor.[40]

Despite these shortcomings, by 1940 almost every American city of over twenty-five thousand residents utilized swine to reduce market waste. A USDA study published in 1941 provided extensive information about the nation's urban waste stream and the disposal techniques, which included incineration, burial, fill, reduction, and feeding to hogs. Garbage feeding was most popular in the mountainous western states (Montana, Idaho, Wyoming, Utah, Nevada, Colorado, Arizona, and New Mexico), with almost 85 percent of the region's total garbage output fed to hogs. The west-north-central states (Minnesota, the Dakotas, Nebraska, Iowa, Kansas, and Missouri) fed approximately 77 percent to hogs, while New

**Figure 20**

The steam-powered shovel deposits restaurant waste and garbage on the waiting hogs at Fontana Farms, circa 1935. Source: Los Angeles Public Library Photo Collection. Used with permission.

England followed close behind with 75 percent. The Pacific Coast states (Washington, Oregon, and California) ranked next with 66 percent. The USDA author reported some pockets of garbage feeding in the regions where it was less popular. Northeast New Jersey, for example, had a robust hotel and restaurant trade and lots of food waste. The author also suggested that the relative unimportance of garbage feeding in the southern states may have been due to the fact that immunization for hog cholera was not practiced as widely, which increased the risks.[41]

A review of the literature on garbage feeding conducted in 1951 indicated that it remained popular, especially in California. In 1950, California's Agricultural Extension Service published "The Garbage Hog Feeding Business in California," which stated that the industry grew "Topsy-like" in the late 1940s. In 1947, there were 355 garbage-feeding operations in the state and 417 in 1948. In Southern California, garbage feeders raised 5.5 million pounds of pork from 1945 to 1948, with fifty pounds of garbage required to make one pound of pork. The state Crop and Livestock Reporting Service estimated that approximately 40 percent of the state's hogs were garbage fed.[42]

The authors of the California study acknowledged that the practice of garbage feeding was changing. The biggest change was the move toward cooking waste food before using it as livestock feed. Uncooked waste often contained raw or undercooked pork from commercial kitchens, which could carry trichinae. Garbage-fed animals could ingest the trichina parasite, which could be ingested by consumers. A 1951 report concluded, "The evidence is overwhelmingly in support of the view that the hog fed on uncooked garbage is the chief source of human trichinosis." Cooking was the answer. Waste cooked to a temperature of 132 degrees Fahrenheit killed trichina larvae, with the considerably higher temperature of 212 degrees for thirty minutes required to kill the hog cholera virus.[43]

Despite the persistence of garbage feeding, concern over the hog cholera threat contributed to a precipitous decline in the practice after World War II. Cooked garbage also threatened to erase the profit margin of garbage feeding operations. In Iowa, for example, the legislature passed a law in 1953 that required cooking all garbage for hogs at the recommended temperature of 212 degrees for a half hour. This measure forced many feeders out of business, their numbers falling from over four hundred in 1953 to just sixty in 1954. In contrast to their California counterparts, however, the Iowa feeders were small-scale operations that used garbage "as a sideline" rather than a major part of the farm.[44]

As garbage feeding faded, there was a significant decline in the number of cases and deaths from trichinosis in the United States. Autopsies conducted from 1931 to 1942 showed that there was a 16 percent rate of infection during that period, while the rate fell to approximately 4 percent by the late 1960s. During the late 1940s and early 1950s, there were over

**Figure 21**

One of the perils of garbage feeding was the transmission of trichinosis. This 1941 diagram depicted the ways in which garbage-fed hogs consumed uncooked, infected pork and spread it to humans. Source: *Meat for Millions: Report of the New York State Trichinosis Commission* (1941). Public domain.

three hundred cases reported yearly, with nine to fifteen deaths per year. The number of cases in the 1960s was much lower, with an average of under three deaths per year that decade, excepting a surge in 1961. The decline in garbage feeding played a major role in reducing trichinosis in the postwar period, more so than the aggressive campaign to cook pork to the proper temperature.[45]

New regulations, however, did not eliminate feeding food waste. Despite the risk of spreading hog cholera, the reality was that garbage feeding still solved problems. According to one observer in 1972, the disposal of urban food waste was a major problem due to the risk of pollution and health hazards, not to mention the use of increasingly precious landfill space for material that could be "recycled through swine."[46] Research continued on the value of food waste as feed, with a 2004 University of Florida Institute of Food and Agricultural Sciences report noting that despite the decline, many states continued to practice it. Food waste from restaurants, military bases, hospitals, nursing homes, and casinos utilized on New Jersey farms averaged over 20 percent crude protein and

25 percent fat, albeit with low fiber and mineral content. Food waste, then, remained a valuable pig feed when accompanied by supplements.[47]

### PIGS VERSUS PEOPLE: ODOR AND SUBURBAN GROWTH

Postwar garbage feeders cold not resolve their odor problem, however. The foul smell of garbage feeding operations was more potent than that of the actual hog feces. During the nineteenth and early twentieth centuries, Americans pushed pigs and garbage feeding out of the central city to the margins, but that was precisely where people were moving in ever-growing numbers during the postwar period. Middle-class Americans sought the clean air, space, quiet, and status of the suburbs, just as their middle-class predecessors had. The difference was that after World War II the middle class was larger and expanding rapidly, not only in numbers but also across space. Suburbs were the new normal for these Americans with automobiles. The new suburbs, however, were located on the same margin occupied by garbage-fed swine, pitting the interests of garbage feeders versus those of the new suburban residents from coast to coast.

Although Los Angeles had solved its garbage problem in the 1920s, the city was in trouble after World War II. Fontana Farms's garbage feeding operation that reduced 470 tons of Los Angeles waste per day announced that it would cease operations in 1951. At its peak in the 1940s, Fontana Farms fattened sixty thousand hogs per year. With the announcement of Fontana's closure, LA's neighboring counties of San Bernardino, Riverside, and Orange all quickly adopted ordinances limiting the size of hog farms to well below Fontana's capacity, blocking the city from an easy, inexpensive waste disposal solution. Yet, as the president of Fontana Farms conceded, "Any livestock has an odor. People don't want it right under their nose."[48]

When one operator obtained a permit to feed LA waste to ten thousand hogs in Haskell Canyon, the *Los Angeles Times* reported that opponents "screamed to the heavens" about the transport of garbage through their communities, complaining of odor and flies. Antelope Valley residents opposed an application for a fifty-thousand-head feeding operation in their community. According to one journalist, "Antelope Valley residents all the way past Palmdale and Lancaster are screaming about being converted into the county's 'slop bucket.'"⁴⁹ Garbage feeding promoters promised that smaller operations could be located in remote canyons where, coupled with modern production techniques, the smell would be of minor concern, but even so residents did not want the waste or the pigs.

The people of Connecticut also experienced the tension between rural people and their pigs and the new suburban neighbors. In a foreshadowing of conflicts to come, residents along Ridge Road in Wethersfield complained of the "abominable, nauseating, and indescribable" odors from nearby piggeries in midsummer 1942. State and city health officials labeled the piggeries a nuisance and required the farmers to clean up their operations and obtain garbage collection permits to stay in business.⁵⁰ Windsor residents made similar complaints in 1945. As many as ten pig farms that relied on Hartford garbage were located there. Hartford's waste often remained in pens for several days, creating such a strong odor on a hot day that people could smell it a mile away. The city building inspector emphasized that the pigs were less of a problem than the garbage. One Hartford resident complained that it was disgraceful that the city endured "vile-smelling piggeries on our outskirts," neglecting to mention the offense to neighboring Windsor.⁵¹

Complaints persisted in Connecticut throughout the 1950s. From 1950 to 1952, residents of Manchester objected to a Hillstown Road piggery operated by Samuel and John Lombardo, who held a contract with the town for refuse removal. The Lombardo brothers were actually wholesale meat dealers who obtained the garbage contract in 1950, but they opened the piggery without authorization from the town. In 1952, the town manager issued a cease and desist order for feeding garbage at the Lombardo piggery. The town proposed to dispose of the waste at the unused sludge beds at the nearby sewage treatment plant or at the town dump where it had been disposed of prior to 1950. The Lombardo brothers pledged to fight the order.⁵²

Warm weather invariably brought renewed complaints about Lombardo's Hillstown Road operation. Following a June 1958 inspection, the town manager ordered the Lombardos to stop dumping garbage until they cleaned up the farm. In August, town leaders required all garbage feeders to cover waste with soil before sundown to reduce smell. Once again, the Lombardo brothers cleaned up until fresh complaints of foul odors in June 1960 brought town officials out to Hillstown Road, who gave Lombardo "a clean bill-of-health." Just one year later, however, Manchester's health director, Dr. Nicholas Marzialo, ordered the Lombardos to cease after an inspection revealed feeding of uncooked garbage to the hogs and forty dead hogs on the property. Authorities arrested both of the brothers and charged them with feeding uncooked garbage to swine, ending the saga on Hillstown Road.[53]

The 1954 "pigs versus people" fight on the border of East Hartford and Glastonberry brought two worlds into conflict—the suburban residential retreat and the farm on the edge of the city. Since the late 1920s, John Querido had fed up to a ton and a half of garbage per day to pigs on his farm that straddled both East Hartford and Glastonberry. Nevertheless, in 1952 construction commenced on a new East Hartford housing development adjacent to Querido's farm. Connecticut law required a minimum of three hundred feet between residences and piggeries. For much of 1953, Querido was not in violation of the law because the new houses were vacant. When residents began to move in during 1954, he moved his hog shelters beyond the three-hundred-foot minimum distance. The hogs, however, could still approach the property line, sparking complaints from new residents. He put up signs facing the new development that stated "This is a pig farm," affirming his prior claim to land use. When town officials ruled that the signs violated local zoning laws, Querido installed new signs that read "This is a Pig Farm for Rent" and "Danger, This is a Pig Farm. Keep Away. Huge Ornery Hogs."[54]

East Hartford residents on Westerly Terrace and Green Manor Drive took Querido to court, claiming that he was in violation of the state sanitary code. A judge agreed that the pigs were too close to the residences, ruling that Querido needed to keep his hogs at least one thousand feet from the housing development on the Glastonberry portion of his farm and completely off his East Hartford tract. He dismissed Querido's assertion that the pigs were there first, contending that the prior claim argument would "stifle the natural growth of a community." He ordered a two-thousand-dollar fine for any violation of the court order.[55]

In 1956, the *Hartford Courant* proclaimed "Controversial Pigs Root Again in East Hartford," noting that Querido had erected a fence on the East Hartford portion of his farm less than forty feet from residences on Manor Circle. Querido insisted that the land was his to use as he pleased and that the city was in error to grant building permits for homes within three hundred feet of the piggery. Two years later, in what the paper labeled "round two" of the "rights of pigs" case, Querido requested a modification to the injunction against keeping a piggery on his East Hartford property, but the tide had turned. Querido's piggery days were done.[56] The law and a rising suburban tide overwhelmed resisters. In Connecticut's case of pigs versus people, middle-class suburbanites won.

While the power of prosperous suburbanites was waxing, they did not always prevail. In 2006, host Mike Rowe of TV's *Dirty Jobs* profiled Bob Combs, the proprietor of R. C. Farms, located just north of Las Vegas, where Combs fed casino food waste to hogs. Viewers watched the process unfold: collecting the segregated food waste from the casinos in semitrucks, hauling it to the farm, dumping it into a sorting area to remove any plastics or cutlery, then transporting it to the heating tank where the slop was held at 212 degrees for thirty minutes, and finally feeding the slop to swine. Throughout the program, Rowe and Combs engaged in good-natured banter about each filthy task while recounting Combs's story.

Combs started in 1963, when his family purchased a hog farm and obtained contracts to remove food waste from multiple Las Vegas casinos. When Combs lost all two thousand of his pigs to hog cholera in 1969, he started over. As of 2014, he hauled tons of garbage every day to feed his twenty-five-hundred-head herd, remaining in compliance with the Environmental Protection Agency's concentrated animal feeding operation regulations for manure runoff and groundwater contamination. It was, by all accounts, a successful business, although as one Combs supporter wrote in 2014, the difficulty for some of his neighbors was the smell.[57]

There was no way to hide the stench, especially because the farm was located in the middle of a suburb. The city of North Las Vegas expanded and eventually surrounded R. C. Farms. In 2006, a buyer purchased a home approximately 1.7 miles from Combs's piggery and could smell the hogs and garbage. Allegedly, the developer had not provided a disclosure of the location of the farm as required by North Las Vegas. According to a lawsuit filed against the developer, the homeowner could not be in the home without gagging. In a 2013 letter to the editor of the *Las Vegas Sun*, one observer reminded residents, "Bob [Combs] has been in that location for more years than I remember. People who bought a house in that area should have done their homework."[58] John Querido may have lost his fight in Connecticut in the 1950s, but as of 2014 R. C. Farms was an ongoing concern, along with other farms operated by Bob's sons, busily converting tons of food waste into flesh while also generating a powerful odor. Combs sold his property to developers for twenty-three million dollars in 2016, retiring on his own terms and benefitting from the development that made his piggery an island in suburbia.[59]

Bob Combs's piggery was a reminder that city dwellers continue to utilize swine for more than just food. Since the first settler cities emerged in America, people and pigs have found distinctive and complementary ecological niches in the chaos. Omnivorous, free-range hogs have scavenged for waste and converted garbage into pork, a good bargain for the working poor. As reform-dominated city councils removed free-range pigs from the streets, however, piggeries on the periphery became more important, providing waste reduction and producing hogs for local markets. The garbage-fed pork that Americans raised in the edge habitat between city and country was an important part of the American economy. By the 1950s, however, Americans had increasingly framed the issue as pigs versus people, a conflict in which city dwellers marginalized garbage feeding in physical as well as social and economic space.

CHAPTER 5

# TO MARKET, TO MARKET

If pork was a vital part of the American economic and dietary landscape, how did Americans transform pigs into pork and other products? How were swine and meat marketed, and who were the ultimate consumers? The production and marketing of hogs, barreled pork, lard, and bacon reveal the ways in which Americans organized their economy as well as the interplay between taste and cost, domestic and foreign demand, and profit and loss.

Pork, lard, and by-products were constant components of the American economy, even in the midst of changing tastes and demand. American pork was a critical part of the colonial food supply for the mainland settlements, but it was more important as an export commodity. The expanding empire meant ever-longer drives to market for producers, sometimes across the Appalachian Mountains. The rapid growth of corn and hog culture in the emerging midcontinent Corn Belt provided the raw materials for a massive hog-killing enterprise in the region centered first on the Ohio River and later in Chicago and the Upper Midwest. American pork and lard fed enslaved people, farmers, and urban workers, not to mention the industrial workers of Britain and Europe. The lard literally greased the engines of American industry and provided illumination through dark nights. Wartime demand invariably resulted in higher prices, booming production, and, at war's end, an unknown future in which the pork industry attempted to encourage demand and maintain market position in relation to other products.

## PORK AND LARD IN THE ATLANTIC MARKETPLACE

Pork was an ideal commodity for the colonial world. Readily preserved with salt and brine, it was far more appealing to English tastes than salted beef or mutton. As a result, wherever the English went there was preserved American pork. Colonial American pork was a major export to the Caribbean sugar colonies, to other English colonies as far north as Newfoundland, and for sailors of the Royal Navy and the merchant fleet. British sea captains relied on Hawaiian hogs for the resupply of ship stores. George Vancouver and James Cook obtained almost thirteen hundred pigs during their stops in the islands. Many of these animals were likely gifts, but their presence indicates significant hog populations in Hawaii. Although the staples of tobacco and rice were far more profitable and significant in the global economy, pork and lard were necessary commodities consumed at home and abroad.[1]

Colonists understood the global demand for protein and the American role in meeting that demand. Samuel Maverick, a settler in the Boston area, remarked in 1660 that pork was critical to the New England economy. "And withal to consider," Maverick pondered, "how many thousand neat beasts and hogs are yearly killed and so have been for many years past for provision in the country and sent abroad to supply Newfoundland, Barbados, Jamaica and other places." Englishman John Josselyn traveled to New England in the late seventeenth century and commented that the merchant "buys Beef, Pork, Peafe, Wheat, and Indian Corn, and sells it again many times to the fishermen." By 1775, the author of *American Husbandry* claimed, "a considerable export from New England constantly goes on in barreled pork."[2]

Almost every English colony boasted of its importance as a pork provisioner. One observer wrote of New York in 1716 that "the Chief produce of this province is beef, flour, pork, butter, and cheese, which they send to the West Indies, and sometimes to Lisbon." Market-oriented farmers in Rhode Island saw opportunity in the West Indies trade as early as the 1670s, with both home-packed and butcher-packed pork from Newport exported to Barbados, Nova Scotia, and Newfoundland. A visitor to Pennsylvania in 1744 claimed, "The staple of this province is bread, flower [sic], and pork." While Virginians relied on tobacco in the seventeenth century, Andrew Burnaby reported that colonists produced "also for the Madeiras, the Streights, and the West-Indies, several articles, such as grain, pork, lumber, and cyder."[3] A study of early eighteenth-century farms and plantations in Talbot County, Maryland, indicated the importance of pork as a commodity. As historian Paul Clemens stated, "As the slave-labor force grew, so did the hog population; by the 1730s Talbot had enough hogs to supply the 7,000 county residents with pork and bacon and a healthy surplus of livestock products for the export trade."[4]

There were numerous brokers in the colonial meat economy. Jonathan Trumbull of Connecticut (later colonial governor) was a hog buyer who butchered hogs and packed the salted pork in barrels for export. In Virginia, enslaved people on Carter Burwell's plantation produced significant pork surpluses, with Burwell marketing as much as ten thousand pounds per year in the provincial capital of Williamsburg during the mid-eighteenth century. Locals ate some of the meat from the Burwell plantation, but more of it was resold as ships' provisions or to sugar plantations in the West Indies.[5] The most famous of colonial pork packers and exporters was the Pynchon family of the Connecticut Valley. John Pynchon was a fur trader during the 1630s and 1640s, but shifted his efforts to farm products by the 1650s. The Connecticut Valley was exceptionally good farmland, and farmers there soon produced such diverse products as onions, grain, horses, cattle, and hogs. Massachusetts farmers along the Connecticut also brought surplus pork to Pynchon and sometimes paid their rent in pork. Pynchon subsequently exported pork.[6]

The French settlements in the Illinois Country carried on a significant provisioning business for the sugar colony of Louisiana. In 1732, Illinois farmers shipped approximately two thousand pounds of beef and between six and seven hundred hams to New Orleans. John Dodge of Illinois shipped ten thousand pounds of bacon and hams to New Orleans in 1788. Like the English colonies that provided wheat, rice, and meat to sustain the plantations of Jamaica, the French developed their own New World dependency.[7]

## HOG DRIVES AND DROVERS

While the French enjoyed easy access to markets along the Mississippi, access was a challenge for many others, especially in the West. Rather than ship bulky preserved meat, many Americans sent animals to market on the hoof. Hog drives of varying lengths were common across the United States in the early republic. Hogs traveled between eight and ten miles per day in droves that ranged from a few dozen to several thousand head crossing creeks at shallow fords and rivers by ferry. Drovers purchased hogs from farmers who had just a few animals as well as those engaged in large-scale fattening.[8] Ohio farmers engaged in the droving business in the early 1800s, supplying Detroit with animals that were processed and sent to Montreal. By the 1810s, Ohioans began to drive hogs and cattle by the thousands through Virginia to Richmond, Philadelphia, Baltimore, and even New York City.[9]

One of the most colorful accounts of the trade is that of Oliver Johnson, recorded by his grandson and published in 1951. Johnson recalled the drives of the 1830s and 1840s that took place in the White River country of Indiana. Two generations removed from the hog drives, the younger Johnson could not have told the story exactly the way his grandfather had, but the account is consistent with other sources. Johnson noted how several farmers would "club together on a drive," to gather a herd of several hundred animals, with each farmer's pigs marked with distinct ear notches. A boss on horseback led each drive, with six to eight young men as drovers on foot and a horse-drawn wagon "to pick up the hogs that give out."[10]

Conditions were difficult for both men and hogs during the fifteen- to twenty-day trek from Indiana to Cincinnati. When a pig could not go any farther, "three or four of us wasted no time a pickin him up, mud and all, and shoving him in [the wagon] till he got rested up." A full wagon meant traveling ahead to the day's stopping point, unloading, and then returning to collect more stragglers. At some point in the late afternoon, the tavern keeper would ride up "to find out from the boss how many men to expect, how many hogs, and how much corn to put out for the night feed."[11] Arriving at the tavern, the drovers penned the herd, unloaded tired hogs from the wagon, fed and groomed the horses, and cleaned up. In Johnson's account, cleanup for drovers meant taking turns standing in a tub of water and being scrubbed with a broom. "After a slug or two of whisky," he explained, "we had a supper of hot biscuits with honey or maple syrup and a slab of ham" and then to bed. The next day the boss settled accounts with the innkeeper for corn, room, and board, and the process began again, regardless of weather.[12]

Historian Lewis Gray reported the tremendous traffic in swine that moved through the Cumberland Gap and other points during the autumn. "In 1836," Gray explained, "the movement of hogs alone from Kentucky was reported as 82,000 through Cumberland Gap, 60,000 on the Kanawha route, and 40,000 through Tennessee to Alabama and Georgia."[13] A Tennessee settler wrote in 1816 that land speculators also engaged in fattening hogs and driving them to Georgia and South Carolina. Settlers along the French Broad River in Tennessee and North Carolina drove hogs across the mountains. One settler recalled that during the 1850s, "in October, November, and December there was an almost continuous string of hogs from Paint Rock [Tennessee] to Asheville [North Carolina]." Such long drives were common throughout the nineteenth century, allowing farmers in the interior to earn the cash needed for essential goods, taxes, and mortgage payments.[14]

Hog drives promised profit but posed significant risks. One Illinois hog drive from the Sugar Creek settlement to American Bottom in December 1836 ended in disaster. The weather turned bitterly cold, forcing the drovers to seek shelter in a cabin and the herd of a thousand hogs into a pile for warmth. The pigs on the bottom of the pile were smothered, while those on the outside froze to death, resulting in a complete loss for the men who assembled the herd. In the Iowa Territory in 1842, a group of settlers responded to a request for bids for fifteen thousand pounds of pork from the Winnebago Indian Mission, located near Fort Atkinson. A group of six men embarked on a hog drive in knee-deep snow in mid-December. Three ox-drawn sleds broke a path for the herd of 125 hogs. They covered fifty miles in eight days to reach the mission, where they slaughtered and butchered the hogs over the next five days. During their return trip the drovers nearly died in a blizzard, compelling them to dig snow shelters before they encountered a group of traders who shared supplies and transported them to safety. Fluctuation in the price of hogs was also a major risk, with a flooded market bringing ruinously low prices.[15]

The destination for many hog drives was the river towns of Ohio, Indiana, Kentucky, Illinois, Iowa, and Missouri. It was a simple proposition to ship pork south to New Orleans from river points such as Cincinnati, Ohio; Madison and Terre Haute, Indiana; Louisville, Kentucky; or St. Louis, Missouri. Barreled pork moved east via the Ohio River and Pennsylvania Main Line Canal, or north and east through an Ohio canal to Lake Erie and then on the Erie Canal to the Hudson River and New York, the largest port in the United States. The packing establishments in the river towns attracted farmers in search of markets. An Englishman described traveling across southern Ohio in the 1850s this way: "The whole country . . . seemed to be swarming with pigs, and long trains of trucks filled with them were pouring into 'Porkopolis' [Cincinnati], where upwards of half a million are slaughtered in the autumn."[16]

Ultimately, railroads and trucks supplanted the hog drives as the principal means of getting hogs to market. By the late nineteenth century, the railroad network reached virtually everywhere in the major hog-producing regions. While farmers still drove animals to town where the buyers representing the packers operated, hogs could be loaded and shipped to urban packing centers with significantly less shrinkage than by driving them overland. As the packing industry fragmented after World War II, trucking replaced rail travel. Packers located in rural areas and in small towns could pay workers less and worry less about pollution, all while taking advantage of the new interstate highway system for moving packed pork to market.[17]

## PORKOPOLIS

In the early decades of the nineteenth century, the center of the pork trade moved westward with the settlement frontier. Businessmen and investors in Ohio, Indiana, and Kentucky quickly became packing leaders, taking advantage of their access to excellent river transportation, cold winters, seasonal surplus labor, and proximity to corn-based hog farming that characterized western agriculture. The easy river access was seasonal, but cold winters also gave the river cities an advantage as packing towns. When the rivers froze, merchants had already sold their goods to country merchants and had empty warehouses, which were available as sites for slaughter and butchering.

The seasonal labor market was critical for western river towns to become packing centers. Once the harvest was completed and surplus livestock

sold, there was simply less physical labor needed on the farm until spring. Entrepreneurial packers recruited employees from the rapidly growing farm population. An older son or father who knew his way around a hog carcass could find winter work in the packing business. Men could return home from the city in time for spring work with cash in hand.

The growth of hog raising further facilitated successful packing in the West. Thanks to consistent annual rainfall over thirty inches, long and hot summer days, and a growing season of at least one hundred frost-free days, there was plenty of corn, more so than on farms of the Northeast. Plentiful corn meant inexpensive hog feed. While corn was the most important crop across the South in terms of acreage, southerners committed large tracts to cotton and other nonfood staples, which meant that southern corn was more important as feed for draft animals and people than as hog feed.

Pork packers needed high-quality salt. For every two hundred pounds of packed meat, as much as fifty pounds of dry salt was required to preserve it, plus the salt to make the brine for barreled pork. There was, however, little quality salt available in Ohio during the 1810s. Locally produced salt from Kentucky, Ohio, and Virginia was high in lime and magnesium, which resulted in inconsistent or failed pork curing. Superior imported salt was expensive to bring across the Appalachians prior to the opening of the Erie Canal or upriver prior to steam navigation. By the

**Figure 22**

This 1868 engraving from *Harper's Weekly* shows a Chicago-bound western hog drive. Driving hogs to market was preferable to butchering and packing market pork on the farm. It was easier to let animals walk to market than to incur the costs of shipping bulky pork and risk spoilage. Drovers on foot, on horseback, and in buggies or wagons kept the animals moving up to ten miles per day before bedding and feeding them at a tavern or drove stand. Source: HarpWeek. Used with permission.

**Figure 23**
Shipping hogs to market by rail was common from the late nineteenth century through the mid-twentieth. At small towns buyers assembled lots to ship to the meatpacking plants in urban areas. Source: Ag Illustrated. Used with permission.

1820s, packers in Ohio, Kentucky, and Indiana took advantage of improved transportation and mixed the lower quality "soft" salt of Virginia's Kanawha River Valley with imported, high-quality "hard" salt, which made reliable pork processing possible. Without this salt, there would have been no western packing industry.[18]

The country merchants who accepted farm-packed pork often became the first commercial pork packers because they possessed the cash and access to the considerable credit requirements to conduct packing operations. As historian Margaret Walsh noted, start-up capital requirements were low for tools and pens, but the cash needed to conduct operations was considerable. While a merchant of the 1810s and 1820s might pack dozens or hundreds of animals, the larger firms of the 1830s and 1840s processed tens of thousands. By the 1850s, packers required from $11,000 to $180,000 for labor, animals, barrels, and salt during a packing season.[19]

Cincinnati became the leader among American packing towns, acquiring the nickname "Porkopolis" in the 1820s. George W. Jones, director of the Cincinnati branch of the Bank of the United States, regularly boasted of the city's rising reputation as a pork packer. One of Jones's correspondents in Liverpool had someone create two papier-mâché hogs labeled "George W. Jones, as the Worthy Representative of Porkopolis" and sent

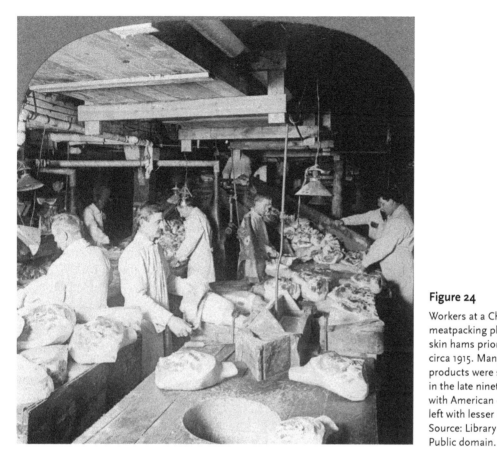

**Figure 24**

Workers at a Chicago meatpacking plant trim and skin hams prior to pickling, circa 1915. Many of the best pork products were shipped abroad in the late nineteenth century, with American consumers often left with lesser quality meat. Source: Library of Congress. Public domain.

them to Jones, who likely spread the story of the good-natured joke.[20] By 1835, a Cincinnati merchant rightfully proclaimed that Cincinnati was the greatest pork market in the world, proving the reality of the Porkopolis moniker. "The number of hogs slaughtered annually, and the perfection and science to which the art of 'hog-killing' has been brought, is indeed astonishing."[21]

Visitors to Cincinnati inevitably commented on the ubiquity of pigs and pork in the city. "Cincinnati," stated Isabella Bird in 1854, "is the city of pigs."[22] A Virginia visitor in 1836 claimed that it was impossible to avoid swine in Cincinnati. "Read their daily papers," he noted, "and every column is filled with 'Pork for Sale' [or] '100,000 pounds of Pork for Sale' . . . you see at every step 'Pork House.'"[23] A Russian visitor, Alexander Lakier, noted that he encountered pigs everywhere in the crowded city streets during his 1857 visit.[24] Most famously, Frances Trollope called attention to the place of hogs in the city, complaining that during a city stroll "the chances were five hundred to one against . . . reaching the shady side without brushing by a snout" and that outings were "curtailed in several directions by my old Cincinnati enemies, the pigs."[25]

Swine arrived in Cincinnati and other packing towns during the late fall to meet their deaths. The animals were gathered in pens next to the slaughterhouses, often in lots of fourteen to fifty animals. Slaughter, butchering, and packing commenced when temperatures reached freezing. A man with a three- or four-pound sledgehammer waded into the pen and struck each animal in the head. Other men drug each stunned animal into the slaughterhouse, where yet another man would stick the neck with a long, thin knife to sever the jugular vein. The carcass was immersed in a vat of hot water, pulled out, and placed on a large table where several men with iron scrapers removed and collected the loosened bristles. They hung the carcass by the hind legs and then "dressed" it by cutting open the body cavity, gutting, washing, and hanging it overnight to cool. As one English observer summarized in 1855, "from the moment piggy gets his first blow till his carcass is curing . . . not more than five minutes elapse."[26]

The next day the cutting began. Frederick Law Olmsted observed the scene in the 1850s and described it as a "human chopping-machine": "A plank table, two men to lift and turn, two to wield the cleavers, were its component parts. No iron cog-wheels could work with more regular motion. Plump falls the hog upon the table, chop, chop; chop, chop; chop, chop, fall the cleavers. All is over. But, before you can say so, plump, chop, chop; chop, chop; chop, chop, sounds again. There is no pause for admiration."[27] Workers threw each cut piece to the trimmer's table, where other men trimmed it and pushed it through a hole in the floor to the meat cellar for sorting, salting, and packing.

Barreled pork was mixed with alum and salt and filled with brine. Each barrel contained two hundred pounds of pork, fifty pounds of salt, and brine solution weighing another hundred pounds or so. Hams were taken to smoke rooms, where they were hung for several months. Meanwhile, all of the lard had been collected, chopped into pieces, and taken to the rendering kettles. Small bits of lard were valuable for sausage making and were collected separately and sold.[28]

During the 1840s, the distinct businesses of slaughtering and packing merged into consolidated and large-scale enterprises. A Scottish visitor described the new organization and scale of the pork-making business in 1854: "The largest of these establishments," he observed, "consists of a series of brick buildings which cover nearly two acres. Here an inclined plane leads from the ground to the top of a house four stories high, and along this the hogs are driven to an upper floor to be slaughtered, and where as many as 4000 can be accommodated at a time." Pork packers learned from the grain millers, who hoisted grain to the top floor for milling and let gravity help convey the milled grain toward the first floor for bagging. At the packinghouse, hogs journeyed to the top floor under their own power and gravity conveyed the carcass, cuts, entrails, and lard downward.[29 w]

The introduction of the horizontal wheel in the 1850s was one of the major technological developments that facilitated rapid pork processing. A ten-foot-diameter horizontal wheel was positioned six feet above the floor with eight large hooks placed four feet apart to lift and move the scraped hogs to stations without workers manually lifting the carcass. As soon as the hogs were scraped, the men inserted a gambrel between the two back legs, which was placed over one of the hooks. The wheel rotated and the carcass moved to the next station where any remaining hair was removed and the carcass washed, pausing twenty seconds at each station. Another turn of the wheel carried the carcass for gutting and final rinsing before the carcass was removed and carried to the cooling room.[30]

The overhead railway loop further automated Cincinnati packinghouses by 1862. James Wilson and Benjamin Eggleston are credited with installing a suspended track system that carried the scraped carcass to workers and into the cooling room, where men used a large iron fork to push the gambrel off the track and onto the cooling hooks. The rapid spread of the railway system developed at Wilson & Eggleston Company reduced the manpower requirements for pork packing, allowing Cincinnati packers to cut costs.[31]

Even as Cincinnati packers mechanized, Chicago supplanted Cincinnati as the country's leading pork packing city. As historian William Cronon described in *Nature's Metropolis: Chicago and the Great West*, there were many reasons for the shift that had to do with Chicago's developing relationship with the hinterland, especially the growth in railroads. Yet

**Figure 25**

This chromolithograph produced by the Cincinnati Pork Packers' Association for the Vienna Exposition of 1873 depicts pork packing in Cincinnati, circa 1873, from shackling the leg prior to killing to salting. The image depicts an industrial triumph, with overhead conveyors, massive rendering tanks, and an army of laborers employed in transforming pigs into commodities. Source: Library of Congress. Public domain.

Killing.

Cutting.

Salting.

IG IN CINCINNATI.

the boom in railroad construction initially helped the packers of the Ohio Valley, not Chicago. Feeder lines into cities such as Cincinnati and Indianapolis allowed farmers from greater distances to get animals to market more efficiently with less shrinkage through the 1840s. By 1850, almost all the hogs that arrived in those cities came by rail. Nevertheless, as historian Doug Hurt noted, Ohio railroad debt meant increased taxes for property owners, especially farmers. Higher taxes served as an inducement for farmers to grow more lucrative cash crops such as wheat rather than raise hogs. Furthermore, Cincinnatians were so committed to the canal and river trade that they were slower to invest in railroads than cities farther west in Illinois and Iowa. By the time there was interest in making additional investments in railroads, the financial panic in 1857 sapped local resources, while Chicago packing had already begun to surge.[32]

The traditional emphasis on packing in Cincinnati and Chicago obscures the fact that many other cities were important packing centers, providing farmers with local markets. In New York, Rochester and Batavia were important regional packing centers. Cheap western corn, however, made it difficult for easterners to compete on a large scale. By the mid-nineteenth century, the number of hogs in New York's Genesee Country was in decline as pork packing in the Upper Mississippi Valley was in ascendance. In Wisconsin, packers in the interior towns and cities butchered to meet local demand or to provision lumber camps, while Milwaukee emerged as a major hog processor. By the 1850s, St. Louis, Missouri; Peoria, Quincy, and Beardstown, Illinois; and Keokuk and Burlington, Iowa, processed ever larger numbers of hogs each year.[33]

## MAKING MARKETS

While most American farmers raised, slaughtered, butchered, and preserved much of their own pork, there were significant surpluses that were either home packed for trade at local stores or driven on the hoof to market. Northeastern farmers consumed most of their pigs at home, although there were some notable hog producers who raised large surpluses that reached international markets. Western hogs and pork reached distant markets in the United States on the hoof or home packed and were consumed domestically and abroad. Southern hogs were usually consumed locally, yet even there migrations occurred within the region, most notably from Kentucky and Tennessee into the Deep South.

Who purchased all of this pork and lard? Where did it go? The answers to these questions are as variable as the distinct economies that developed across the country. Small-town and country merchants often accepted pork to settle accounts and even advertised that they would purchase packed pork, which they would resell in larger markets. Samuel Rex of Schaefferstown, Pennsylvania, regularly issued calls for farmers to bring in pork, lard, butter, and other commodities in the 1790s and early 1800s. Rex, in turn, sold these items to other customers, many of whom were workers in the growing iron industry.[34] Some producers found niche markets. A planter on Wolf Island, Missouri, situated in the Mississippi River below the confluence of the Ohio and Mississippi Rivers, raised cattle and hogs for the supply of steamboats.[35] Settlers in Dallas County, Iowa, in the 1850s slaughtered and packed over three thousand head per year for sale to western migrants who were rapidly filling and crossing the state.[36] Much of this home-packed pork, however, was of uneven quality.

Some optimistic settlers who hoped to enter the pork trade were disappointed. Managers of the Holland Land Company hoped to exploit hog production in western New York by packing and selling pork in the early 1820s. They failed to do so, however, due in part to some production problems. The company used poor-quality salt, which resulted in spoilage, while the locally made barrels were not tight enough to hold the brine. The marketing program collapsed in 1826, leaving the regional, national, and global pork trade to westerners.[37]

The larger scale packers of Cincinnati, Chicago, and the other meat-packing towns prepared products to suit different market demands. "Bacon" was a generic term for salted and smoked meat, often with the bone left in, although it was possible to purchase sides of bacon. Hams were also packed separately, but most preserved pork was packed using three grades: clear, mess, and prime. Clear pork described the best meat, made from the fattest animals. A two-hundred-pound hogshead of clear pork included four- to five-pound chunks of every part of the animal except hams and ribs and was generally marketed in Europe. Mess pork included the ribs and two rumps in addition to the smaller chunks and was sold for use in the world's merchant and naval fleets. Prime pork was the lowest quality, made from the leanest animals. Each two-hundred-pound barrel of prime pork was made up of two shoulders, two jowls, and sides and was purchased to feed enslaved and free people around the world.[38]

Western pork dominated eastern markets after the completion of the Erie Canal in 1825. As one observer noted in his remarks to the Farmer's Society of Florida, New York, in 1828, "We cannot advantageously make pork for the Albany or New-York markets. The farmers to the West, where Indian corn is cheap, can now, assisted by the canal, afford to undersell us."[39] Ohio pork packed in Sandusky, Cleveland, Toledo, and numerous other Ohio towns located either on the shores of Lake Erie or along the Ohio Canal, flowed into the Erie Canal bound for Albany and New York City, where it was either consumed or reshipped in the coastwise or international trade. In some cases, local investors purchased hogs from farmers, fattened them, and then packed them for the eastern trade. In Bellefontaine, Ohio, one such operator obtained and fattened twenty thousand hogs per year for the eastern trade in the 1850s.[40]

By the 1830s, however, the bulk of the western pork trade reoriented southward to New Orleans. Pork and bacon receipts at Portsmouth on the Ohio River increased from sixteen million pounds in 1839 to seventy million in 1844, more than double the amount that arrived in Cleveland to journey by lake and canal to New York. Ohio and Indiana pork was increasingly bound for New Orleans.[41]

Western pork that arrived in New Orleans ended up at many destinations. Portions of the western pack fed Europeans and urban Americans. One visitor to Cincinnati in 1839 reported that 190,000 hogs had "heaved for the expiring groan and have been packed for the consumption of you Bostonians and your neighbors along the coast."[42] Frederick Law Olmsted concluded his description of the Cincinnati butchering process by stating that "hams, shoulders, clear, mess and prime" were all shunted to different points in the factory, "each to its separate destiny—the ham to Mexico, the loin for Bordeaux."[43] Foreign markets, always important, were growing, especially England during the 1840s and 1850s. According to one Liverpool correspondent, American lard was the most extensive import for cooking, soap, and candle making. Bacon and hams were also popular in the English market. The advice from England was for Americans to emulate Irish pork, noted for firmness. Some of the American pork was "covered with a thick, slimy, red matter," which hurt prices. The English also complained that American hams were oversmoked or too salty. Lard oil found a ready market in England as a lubricant.[44]

The American South was an increasingly significant market for western pork. On the surface, southern purchases of western pork do not make sense, given the fact that southern states were leaders in American hog production through the mid-nineteenth century, with Kentucky and Tennessee consistently ranked among the top states. But, as geographer Sam Bowers Hilliard demonstrated, hog production was not evenly spread across the South. Thanks to the extreme profits in sugar production, Louisiana planters in the sugar parishes did not meet their internal demand for pork. The booming cotton frontier absorbed even more of the pork that arrived in New Orleans. It came to Mobile, Alabama, then traveled on steamboats up the Alabama and Tombigbee Rivers to supply upland cotton plantations. Charleston, South Carolina, was another important market for pork from New Orleans, with some bound for Savannah and Darien, Georgia, to sustain enslaved people on coastal cotton plantations.[45]

Planters who purchased pork and bacon did so based on a calculation that labor invested in commodity production was more profitable than that devoted to subsistence. An English observer who traveled in the South during the 1850s noted that the bacon supply on many plantations "is almost entirely imported from the Northern States." As a result, every year plantation owners dedicated "a large sum . . . to pay the hog-raiser in Ohio," not to mention Kentucky mule breeders and northern investors. A Mississippi informant related that it was not profitable for planters with extensive land and slave holdings to raise hogs, "for it is found to be almost impossible to prevent the negroes stealing and roasting the young pigs."[46]

Planters' concerns about slaves stealing pigs were common enough across the South that many planters eschewed raising their own hogs for that reason. It is difficult to know the extent to which masters' fears were real or a generalized suspicion of theft. Slaves helped themselves sometimes to supplement their meager fare or as an act of resistance. Perhaps most importantly, they may have shared a sense that the products of slave labor belonged to them.[47] Many planters inflicted severe punishments on their slaves for pig theft, real or imagined. A Mississippi planter who wished to be self-sufficient in pork production encouraged his overseer to handcuff several slaves who were suspected of nightly pig stealing to stop the thefts. In 1860, a Mississippi planter claimed that he would purchase Ohio pork because he disliked whipping his slaves so often for stealing pigs.[48]

Despite the theft, many large planters produced significant quantities of pork for consumption on the plantation rather than for sale. The slaves on the four plantations owned by David R. Williams of South Carolina raised enough corn and pigs to keep themselves and their owners supplied with pork. On the Telfair Plantation in Jefferson County, Georgia, the overseer reported in 1840 that the pork yield would cover eleven months of rations at the rate of one thousand fifty pounds per month. A Savannah merchant was a reliable buyer for the balance. Alabamian James Mallory reported that he purchased pork for his plantation only one year between 1843 and 1860.[49]

Lard emerged as a major American commodity during the early nineteenth century, one that had value equal to or greater than pork, depending on the quality. Lard was America's most widely used cooking fat in the preparation of fried meats, breads, greens, and vegetables as well as in baking. Trimmed fat was a critical component of headcheese, scrapple, and sausage. These meats lasted up to a year in crocks packed with melted fat, which made an airtight seal to prevent decay. Rural families used lard to make their own soap, heating and combining it with lye leached from wood ashes to make either hard soap cakes or soft liquid soap for cleaning bodies, dishes, and clothing. Manufacturers frequently used linseed oil, lard, and tallow for the lubrication of moving parts of steam engines and machines. Lard was also used for illumination. A saucer or tin cup with a wick of cotton cloth or a custom-made tin lamp with a nozzle for the wick provided feeble light, but as one writer in 1842 recalled of his boyhood, the lard lamp was used only on special occasions.[50]

Industries to process by-products into commodities followed pork packing. Two immigrants, William Procter, an English storekeeper and candle maker, and James Gamble, an Irish soap maker, married sisters from Cincinnati and formed a partnership to manufacture soap and candles in 1837.[51] Chandlers and soap makers like Proctor and Gamble saw profit in the rendering and conversion of lard into commodities. The highest quality lard was in the body cavity, known as leaf lard, and was easily cut out during the butchering process and rendered with little waste. Lard from the back and intestines required more effort. Intestines were washed, cut, and placed in a kettle. The cooked impurities, known as cracklings, were skimmed off, leaving the lard to be cooled for storage or immediate use.

Innovations eased and accelerated the process of rendering and refining lard, facilitating wider industrial applications. In May 1843, one report noted that of the 245,000 hogs killed in Cincinnati to date that year, 80,000 had been killed for the express purpose of conversion to lard using a new technique of steam rendering. Animals designated for this process were killed and the hams removed for curing, with the rest of the carcass put in a pressurized tub. Steam was injected into the tub and the lard reduced and drained off, then strained and packed. According to the correspondent for the *Albany Cultivator*, the amount of lard recovered from the total carcass ranged from 55 to 65 percent, with Berkshire and China hogs yielding as much as 70 percent. The leftover bones from this process were clean and in perfect condition for fertilizer manufacture.[52]

By the 1840s, then, lard promoters envisioned a bright future. Some observers labeled swine as the land whale, prairie whale, and Ohio whale, suggesting that lard would rival or displace whale oil as the leading source of fuel for illumination. This language was so evocative that the *Prairie Farmer* featured a cartoon depicting whalers in a horse-drawn prairie schooner harpooning hogs as they would whales. As the *Milwaukee Sentinel* opined, eastern whalers "may spout and blubber as much as they please, [but] the day for using sperm oil and candles in all the West is soon at an end."[53] New England textile manufacturers and river boatmen placed large orders for western lard oil, and at least one source reported that the US Navy agent at Norfolk, Virginia, had ordered some in 1842. Much of the increase in lard exports during the 1840s and 1850s was due to stearin candle manufacture and the desire for cheap lubricants. Newspapers reported demand for American lard oil from France, England, Mexico, India, and Cuba. The outbreak of the Crimean War in 1854 boosted hopes for even greater American lard exports.[54]

But lard for illumination did not fulfill promoters' dreams. There were skeptics even as the lard boom was under way. An editorial writer from Natchez, Mississippi, pronounced Cincinnati lard oils as "a swinish humbug" for failing to provide reliable light. He joked that lard failed to burn any better than "a live pig would have done." He preferred "a lively pig, dipped in turpentine, and lighted at his caudal extremity [tail], to the [lard] lamp!"[55] More significantly, the 1859 the oil strike at Titusville, Pennsylvania, eventually contributed to a decline in both whale and lard oil for illumination.[56] Cheap and plentiful kerosene crowded lard out of the market. While urban residents increasingly lit their homes with gas and electricity, kerosene was the illumination fuel of choice for the majority of people on America's farms and in small towns after the American Civil War.

The Civil War years were good for northern hog farmers. Despite the loss of southern markets, government buying and robust overseas demand, especially for salted meat in England, were a boon. Treasury Department reports indicate that wartime exports of American ham and bacon increased sixfold, lard increased fourfold, and pork increased by half.[57] Such high demand resulted in a net decline in hog numbers in the northern states from seventeen million in 1860 to thirteen million in early 1865, further supporting prices.[58] High profits in pigs and pork in the loyal states encouraged speculation by commission merchants who purchased packed pork to extract the largest possible payments from the US Army, the single most important pork purchaser in the country. Between late 1863 and the spring of 1864, the price of mess pork had nearly doubled.

The price pressure resulted in the Great Hog Swindle. Major Henry C. Symonds, the depot commissary at Louisville, Kentucky, asked for and received permission from his superiors to begin packing army pork in the fall of 1864 to obtain it at lower cost. Symonds authorized his buyers to pay the current market price of eight cents per pound and then sent them into the countryside, where they purchased several thousand head and drove them to town. Meanwhile, Louisville packers offered over ten cents per pound, infuriating many farmers who already sold to army buyers. Farmers assumed that the army officers were engaged in a fraud to buy at an artificially low price, sell to packers, and pocket the difference. The differential between the country price and the city price, however, was not out of line with common practice. Farmers who sold in the country did not have to incur any droving expenses; army buyers actually paid those costs. Symonds was no crook. Under his authority, the army packed approximately eight million pounds of salt pork that winter and saved the taxpayers an estimated $200,000, but despite Symonds's success in

**Figure 26**

Lard oil as lighting fuel reached new highs as commercial whaling played out during the nineteenth century. This late nineteenth-century trade card from N. K. Fairbanks and Company recognizes the important role of corn in producing hogs with copious lard on the carcass. Source: Historic New England. Used with permission.

packing affordable pork for the government, the incident became known as a swindle. Many Kentuckians were left embittered toward the national government, compounding the problems of sectional reconciliation in a loyal slaveholding state.⁵⁹

The war also accelerated the shift of pork packing from Cincinnati to Chicago. The creation of Chicago Union Stockyards, which opened on Christmas Day 1865, was a turning point. As historian Margaret Walsh claimed, Chicago packers were able to thrive because of the "increased and regular flow of animals" via rail as well as the colocation of stockyards, associated services, and packers, all of which created a centralized livestock market.[60] Cincinnatians, long dependent on the Ohio River, were slow to use railroads for shipping packed pork. Instead, they relied on rail primarily for extending the city's reach into the interior for hog supply. Chicagoans turned to rail transportation much faster, for not only hogs but also transporting packed pork to market. Closure of the Mississippi River in 1861 due to the secession of the Deep South left Cincinnati packers cut off from their primary market of New Orleans.

After the war, packers and farmers in the Upper Midwest developed a dependency based on corn, just as Ohio farmers and packers had in the 1820s and 1830s. As geographer John Hudson noted, the packing plants built at Saint Paul, Minnesota, were removed from much of the hog supply when they were constructed in the 1880s, but by 1900 Minnesota farmers raised half as many hogs as Illinois farmers.[61]

Farmers in the Upper Midwest gained better access to markets with Chicago's ascent. The use of ice for artificial refrigeration in packing plants and rail cars contributed to the success of Upper Midwest packers. Artificial cooling stretched the packing season beyond traditional winter limits into the warm months. Packers hoped that a longer packing season would allow them to invest more in facilities and ride out the low prices of the glutted traditional winter packing season. Chicago packers first utilized ice to pack pork in the summer of 1858. By the 1880s, over half of the annual hog receipts were in July, representing a dramatic shift from the pre–Civil War years when November, December, and January accounted for nearly all the hogs received for packing.[62]

## TRADE WAR, WORLD WAR, AND THE GREAT DEPRESSION

The rise of year-round packing and the boom in hog production lowered costs for American pork in Europe. As a correspondent for the *Irish Farmer's Gazette* lamented in 1863, the "inundation" of the Irish market with

American pork threatened even the Irish cottage pig because "American speculators can furnish the markets at a much cheaper rate."[63] During the 1870s, European crop failures provided a further opening for American corn-fed pork, with exports doubling between 1874 and 1880, accounting for nearly 10 percent of the total value of American pork exports in 1879.[64]

Europe retaliated. In 1879, Italy ordered the exclusion of American pork products. Portugal and Greece enacted their own bans in 1879, and Spain followed in 1880. That year Germany prohibited American chopped pork and sausages and followed with a complete ban on pork in 1883. France banned American pork in 1881 and Austria-Hungary, Turkey, and Romania did so later that year.[65] These bans, however, were promulgated not as protectionism but instead as public health concerns based on the risk of trichina-infested pork from the United States. Trichina, a parasitic roundworm, enters human hosts through the ingestion of uncooked or undercooked pork and causes abdominal pain, diarrhea, fatigue, and fever as the worm bores into the flesh. There was no particular evidence that linked several trichina incidents in Europe to American products, but it was a convenient cover for protectionism and a source of anxiety for American packers. American diplomats recommended a federal inspection law as the best way to undermine European protectionism. Farmers also wanted an inspection law to preserve their business, with expenses paid by the processors or exporters, but the industry opposed it. Packers likely feared that increased costs due to inspection might hurt pork consumption. As historian John Gignilliat observed, "packers and exporters said little about inspection. They preferred to talk about retaliation."[66]

To combat the boycott, President Chester Arthur appointed a special commission to study the trichina problem. The commission, composed of scientists and members of the New York Chamber of Commerce and Chicago Board of Trade, concluded that trichina was present in American herds but to a lesser extent than in European herds. The European trichina problem, according to the commission, was homegrown, not imported.[67] Joseph Nimmo, chief of the US Bureau of Statistics, echoed this finding in a report for the Treasury Department. Nimmo solicited evidence from industry and public health officials to prove that American pork was safe. He asserted that Americans consumed five times the amount of pork than was exported and that there was no significant trichina problem in the United

States. If American herds were infested, he argued, surely Americans would complain and physicians would report the problem. Since US Navy personnel consumed a meat diet consisting mostly of salt pork, Nimmo requested information from the supervising surgeon-general of the Marine Hospital Service to understand the extent of the problem. The surgeon-general replied that in over 234,000 personnel treated at the hospital between 1874 and 1884 there were no reported cases of trichina infection.[68]

The European meat ban was resolved by 1891 through a combination of diplomacy and domestic politics. The major obstacle of protectionism had been resolved in 1890. The McKinley tariff removed the US duty on imported sugar from Germany, easing the way for Germany to admit American pork. The French government raised the duty on American pork and repealed their ban. Other nations quickly followed suit. President Benjamin Harrison signed the Meat Inspection Act in 1891, placating European health concerns. The face-saving law provided for microscopic inspection of all pork processed for export, a measure that the secretary of agriculture had first endorsed in 1889.[69]

In the meantime, Europeans demanded leaner American hogs. In 1898 and 1899, University of Minnesota professor Thomas Shaw attempted to instruct American farmers in a two-part essay published in *Prairie Farmer*. Shaw cited the example of Canadian hog farmers in Ontario who laughed at the idea of a bacon hog in the 1880s, but soon found that those who raised meatier animals earned a premium compared to those who continued to produce heavy animals with lots of lard. In 1897, the Sinclair packing firm noted that English and Danish bacon sold for double the price of American bacon, with the premium going to the product with a "streak of lean."[70] To obtain leaner American bacon, a Michigan farmer urged his compatriots to feed bran and middlings (coarsely ground grain) and to provide pasture. Corn was important, but only in combination with pasture.[71] Similarly, closer attention to management, including housing, farrowing, and sanitation, would allow for the creation of a hog that better matched market demand.

American packers favored lighter hogs due to demand for leaner bacon and the shrinking market for lard. Since the 1830s, American markets had prized the lard hog, "as the taste was for heavy fat pork, large hams, and abundant use of lard," noted a writer for the packer Swift & Company in 1898. The "national taste," however, was changing, with "growing demand for something delicate in the matter of hog product; for light hams and the well-streaked bacon that is furnished by 150 to 175-pound hogs."[72] The American housekeeper, according to one author, had "become somewhat epicurean in choosing her meat, and hence it is that the packing trade to-day finds that course, fat-laden hog products hold no place on the menu of the exacting housewife."[73] As one commentator in the Germantown, Pennsylvania, *Telegraph* observed, "As a rule the sausage sold in our city markets is too fat, and when fried or broiled shrivels to about one-third its original size. If animals of moderate weight were used the products would be more satisfactory to the consumers and more profitable to the sellers."[74]

The continued presence of fatty hogs did not seem to hurt the export economy. Pork exports continued to surge in the early twentieth century, with an estimated 12 percent of American pork and lard exported. Consumers in the United Kingdom, Canada, and Germany ate record amounts of American pork. American lard and lard-based oleomargarine were popular in Cuba, Germany, the Netherlands, France, Norway, and Britain. The mining and manufacturing districts in Germany in particular were the destinations for US cured pork. Only bacon exports suffered, losing ground to leaner Danish bacon in the British market, falling from approximately 80 percent of imports to the United Kingdom in the 1880s to 37 percent by 1910.[75]

World War I boosted demand for American pork. The hog population in England declined 40 percent between 1913 to 1918, while that of France and Italy decreased 26 and 32 percent, respectively. American hog numbers had increased in 1915 and 1916 due to optimistic breeding to meet anticipated European demand.[76] Hogs were critical to the war food supply because the short gestational period and rapid maturation of hogs gave flexibility in the meat supply. Herbert Hoover, war food administrator, hoped to increase pork production in late 1917 through voluntary price increases, but the Big Five packers (Swift, Armour, Morris, Wilson, and Cudahy) were not interested in negotiating. Packer intransigence on prices during a national crisis frustrated Hoover. His chief negotiator observed that after years of claims by packers that 2 percent was the gold standard of profits, they now wanted more. They grudgingly agreed to purchase increased hog numbers at $15.50 per hundredweight in exchange for a government promise to purchase all output at a guaranteed 2.5 percent profit for small packers and 9 percent profit for the Big Five.[77]

The wartime price increase mobilized bureaucrats and farmers. In June 1917, the USDA counseled, "Every breedable sow should be bred to bring a fall litter." Florida's extension director reminded farmers that the nation needed more hogs "at high prices." Farmers who bred four sows the previous year could simply breed one more to obtain a 25 percent increase in income. Raising hogs, he noted, while "man's work," was nothing compared to the challenges "faced by our soldiers in France." Increasing hog production, he insisted, was "profitable patriotism."[78] Even so, the editor of *Wallaces' Farmer and Iowa Homestead* cautioned farmers in 1917 that they would be vulnerable in the event of a sudden contraction in demand.[79]

*Wallaces' Farmer* was prescient. By late 1918, pork surpluses were a source of embarrassment for President Wilson and Hoover. Exports of pork and lard in 1918 had exceeded that of 1917 by one billion pounds. Even more hogs were bound for market in 1919. To absorb the supply, Hoover ordered government purchases of an additional 50 million pounds of pork and lard and millions more pounds for European relief. The United States purchased an additional 250 million pounds for resale to Germany, thanks to congressional appropriations for massive credits to European nations for purchases of American food.[80]

It was not enough to support prices for pork or beef, as both returned to prewar levels in 1921 when European demand lagged. In 1919, European purchases accounted for 21 percent of hog receipts, but that number declined to under 7 percent by 1927.[81] Livestock prices crashed in 1931, with hog prices falling by a third that year and a drop of another 50 percent by 1932. President Franklin Roosevelt's agriculture team developed the Agricultural Adjustment Administration (AAA) program in early 1933 to reduce production and raise prices through voluntary production cuts, but they recognized that farmers might be hesitant to reduce production of feed crops such as corn as long as they had hogs to feed. Cutting corn acres in the spring without reducing hog numbers would result in a fall feed shortage, chilling the impact of the acreage reduction program.[82]

Secretary of Agriculture Henry A. Wallace devised the Emergency Hog Marketing Program in the late summer of 1933 to eliminate a fall swine surplus and ease the way for acreage reduction. To adjust hog production to meet the domestic demand for pork and feed production and to support farm income, the federal government purchased and destroyed six million young pigs, which were converted into fertilizer and grease. The government purchased and slaughtered another two million hogs for food relief.[83]

It was a desperate plan for desperate times. Many people opposed the slaughter of young pigs, questioning how a society could waste food while so many were hungry. Secretary Wallace responded by claiming that he was most interested in easing farmers' financial problems: "I cannot say it too strongly," Wallace argued, "nobody will starve if we reduce hog production; but farmers will go without the necessities of life if we don't." For their part, farmers generally supported the emergency hog-buying program. Iowa farmer Elmer Powers sold hogs in September 1933 and noted that while many farmers he spoke with "regretted taking these fine pigs out of the lot at this time . . . many of these same farmers agreed that something must be done." Financially distressed farmers welcomed the cash.[84]

Most Corn Belt farmers, like Powers, were also receptive to the administration's acreage and livestock reduction program and signed up in large numbers. Powers served on the local AAA corn-hog committee in 1934 and reported high levels of interest in his area for limiting production in exchange for government checks, noting that 120 out of 130 farmers in his township had signed up, including himself. He agreed to plant sixty-five acres of corn, idle sixteen acres, and keep twelve brood sows in hopes of producing fifty-five market hogs. During the subsequent months, Powers reported grumbling at sale barns and general stores about the program, but in early 1935 he claimed that farmers in his area agreed that it was "useful and did them much good." Participation remained high and consequently the population declined from 62.1 million hogs in 1932 to 42.9 million in 1936 and remained low until the war emergency threatened in 1940. Prices improved in 1935 but crashed again in 1939, only improving when the war began.[85]

## FROM WORLD WAR TO POSTWAR MARKETS: THE UK, USSR, USA, AND PRC

World War II pulled American livestock producers out of the depression. In early 1942, Congress passed price control legislation that established support prices for market hogs weighing 180 pounds or more, first at $9.00

per hundredweight and later that year to $10.00. In 1943, the support price was set at $13.75 to boost hog numbers and align production with demand.[86] The USDA Bureau of Agricultural Economics estimated that there were approximately sixty million hogs on American farms in 1940 and seventy million by 1943. As in World War I, "profitable patriotism" was an easy sell, so easy that record hog marketing in late 1943 brought prices down to support levels. In October 1944, Congress reduced the support price to $12.50 per hundredweight as the hog population surged to over eighty million animals.[87] Farmers scrambled to adjust production to expectations of lower postwar demand. By January 1945, government estimates put 1945 production at 7 percent less than 1944, with the heaviest cuts coming from outside the major hog-producing region.[88]

Packers sent massive quantities of pork to American service members and the Allies. For some soldiers and civilians, it was fresh meat, sausages, bacon, and "defense ham," a harder product suitable for rough handling. Spam achieved legendary status in the United Kingdom and the Soviet Union. It was so ubiquitous in England that journalist Edward R. Murrow stated in his 1942 Christmas broadcast from London that, despite widespread privation, "there would be Spam for everyone." Hormel manufactured a traditional Russian dish called *svinaia tushonka*, which consisted of seasoned, lean pork boiled and packed in tins, for export to the Soviet Union. It was so important that Nikita Khrushchev later credited it with helping Soviet forces repel the German army.[89]

The return of peace brought to the United States increased prosperity, which was reflected in changing food habits. Rising incomes meant more beef on the table compared to pork. Beef remained more popular than pork among urban people, and the United States was rapidly urbanizing.[90] Perceptions of pork as unhealthy due to fat content compounded the problem for the pork industry, leading to stagnant demand after the war with some periods of low prices, especially in the mid- to late 1950s.

Pork producers responded with alacrity. To boost consumption, they attempted to reduce the fat content of the carcass and change consumer perceptions. A critical component of this offensive was finding the resources to fund an effort to convince consumers that pork was desirable. As early as 1956, pork producers envisioned creating a fund that would pay for the kind of meat promotion that would resonate with consumers. The National Swine Growers Council advocated for a voluntary program that would permit packers to deduct a few cents from the sale of each hog, known as the checkoff, to fund research and marketing.[91]

The checkoff was a long time coming. In 1966, *National Hog Farmer* magazine published a series titled "Blueprint for Decision" that included two essays on why farmers needed the checkoff. As one leading Iowa hog producer proclaimed, "Let's face facts—we in the corn-hog belt must sell our product in the population centers far from home. We need a national promotion and education arm financed by the producers of many states."[92] A national referendum of pork producers in 1968 indicated sufficient support for a five-cent-per-head voluntary contribution known as "Nickels for Profits." Those nickels ultimately paid for marketing campaigns such as "America, You're Leaning on Pork" and "The Other White Meat." Despite the response in reshaping the hog, per capita pork consumption remained stagnant.

The salvation for American pork producers was the export market. Exports to Canada, Mexico, and Europe continued throughout the postwar period. Canada and Mexico remain the top destinations for American pork, thanks in part to the North American Free Trade Agreement of 1994.[93] Newer markets included Japan, Korea, and China. Just as "The Other White Meat" marketing campaign unfolded, American hog producers began to focus on Japan as a leading candidate for market growth. In an unexpected irony to the American effort to sell pork as white meat, Japanese consumers preferred darker pork. Still, exports to Japan increased in the late 1980s and continued to do so into the 2010s.[94]

The biggest opportunity for export growth was China. By the late 1990s, China consumed approximately half of the world's pork, although none of it was from the United States. High tariffs and complicated health and distribution rules blocked US exports. In 2000, however, the United States ratified a trade agreement that would allow USDA-inspected pork into China in exchange for American support for China's admission to the World Trade Organization.[95] Since then, China has absorbed a rapidly growing percentage of American exports, from less than 1 percent in 2000 to 6.8 percent in 2014. By 2016, exports to China surpassed the volume to Japan, ranking as the third most important export destination for US pork. Chinese dietary preferences are opposite to American tastes, with Chinese consumers favoring fatty cuts and offal and American consumers preferring lean meat.[96]

The Chinese government actively solicited American pork. In 2007, Smithfield Foods negotiated a deal with China to deliver sixty million pounds of pork, and the next year China agreed to purchase just under 5 percent of Smithfield's stock. Surging exports led to closer ties. In 2013, Shuanghui International, a Chinese conglomerate, purchased Smithfield Foods outright for $4.7 billion. It was a stunning move for many American producers. Smithfield now controlled approximately 25 percent of the American hog business.[97] While American pork exports have always played a critical role in the economy, Smithfield's purchase by a foreign entity raised the question of whether Chinese-owned Smithfield pork shipped to China actually constituted an export.

Market access has been critical to the American pork industry, from exports to the Caribbean in the colonial era to the recent opening of China, from the southern cotton frontier to battlefronts around the world. For most of American history, producers utilized low-cost range feeding and grain to produce significant surpluses of pork and lard. These surpluses have often resulted in depressed prices, resulting in European protectionism in the late 1800s. Low prices persisted in the twentieth century, with much of the respite for hog farmers and packers coming during wartime. To stave off losses, pork producers organized and worked to increase domestic demand and exports. Finding domestic demand stagnant, industry hopes focused on the export market. Meat industry insiders have devoted considerable attention to the fact that incomes in developing nations have been rising, fueling global meat consumption that has been favorable to the pork sector, transcending limits on American production.

CHAPTER 6

# SWINE PLAGUES

During the 1850s, a new, virulent plague struck the hog population of the United States, with untold numbers of hogs dead and many more sick and weak. Newspapers in Virginia, western New York, New Jersey, Illinois, and Indiana reported devastating losses. From Princeton, Indiana, a correspondent noted that some farmers lost as much as 80 percent of their herds in 1858. According to the *Ohio Farmer* in 1857, sixty thousand hogs had died of hog cholera within a hundred miles of Cincinnati. Travelers along public roads were subject to the "stench from the rotten carcasses left to moulder and decay . . . and some of the branches are running greasy water, where the carcasses have been thrown in, as a convenient place to get rid of them." In March 1859, a Tennessee farmer named Robert Cartmell wrote that "hog cholera has played sad havoc with hogs throughout the country for months—is nearer me now than I like." Cartmell's pigs did not escape. Hog cholera arrived on his farm in May. By August, he estimated that forty-five of his one hundred thirty hogs were dead.[1]

The label "cholera" seemed appropriate for the dreaded swine disease. Asiatic cholera arrived in the United States in 1832, killing thousands of Americans. In subsequent years thereafter, cholera periodically struck the length and breadth of the United States, killing many thousands more into the 1850s. Traveling from seaboard ports to Great Lakes ports and along American canals, rivers, and roads, Asiatic cholera reduced its human victims to misery, despair, and, all too frequently, death. When a horrifying hog disease appeared in the mid-nineteenth century, cholera provided the most apt comparison, and the name "hog cholera" stuck. Yet "hog cholera" was not the same disease. Like Asiatic cholera, the disease struck quickly and lethally, but Asiatic cholera is a bacterium, spread by consuming drinking water contaminated with fecal matter. Hog cholera (now known as classical swine fever) is a virus spread through oral-nasal contact.

Hog cholera plagued American farmers over the next hundred years, before fading from concern. Government scientists used new instrumentation and techniques to discover the cause by the early twentieth century. The protagonist of this commonly told story is Dr. Marion Dorset, who, along with a dedicated team of USDA scientists and assistants in the Bureau of Animal Industry, discovered the cause of hog cholera in 1903 and developed a serum to control it in 1907. Dorset's discovery ultimately resulted in a sustained eradication effort that achieved success by the late twentieth century.

Today, such a role for the USDA in disease research and prevention appears natural. Nevertheless, this kind of sustained government intervention on behalf of farmers and the meat industry is notable for its novelty. The commitment represented a consensus on the part of elected officials, bureaucrats, and farm interests that American agriculture needed a fix, not only for hog cholera but also for deadly Texas fever in cattle. During the 1880s, when Congress and the USDA first mobilized to deal with livestock epidemics, efforts to control human epidemics paled by comparison. The US government rushed to serve the interests of the meat industry and farmers and to secure the place of both hogs and cattle in the farm landscape.

### HOG CHOLERA: ORIGINS AND CURES

Historical accounts of hog cholera invariably begin in the 1830s, but the origins of the disease remain obscure. As hog cholera became a crisis in the 1850s and remained endemic through the Civil War, Americans understandably began to search for origins. During the 1870s and 1880s,

USDA personnel conducted surveys, asking farmers about their first experiences with the disease. Based on the recollections they gathered, the USDA dated the first occurrence of disease in the United States to 1833 in Ohio, with numerous outbreaks over the next twenty years: ten reports from ten different states in the period up to 1845 and another ninety-five reports from thirteen states prior to 1855.[2]

Yet the evidence of hog cholera in the 1830s is thin. The most reliable report that some epidemic was present during that decade was an observation by Henry W. Ellsworth in his 1840 treatise *The American Swine Breeder*. Ellsworth quoted a correspondent from the *Franklin Farmer*, who observed that there were "new and terrible diseases" affecting swine in the western states.[3] Even so, historian Robert L. Jones, the author of the standard history of Ohio agriculture of the period before 1880, simply stated, "Cholera was not to be a problem till the mid-1850s."[4] More recently, historians Alan Olmstead and Paul Rhode uncovered no contemporary accounts of a disease like hog cholera in Ohio during the 1830s, the reputed place and time of origin. Their conclusion echoed that of Jones. By the late 1850s the disease of unknown origin was present in the Midwest, Northeast, and Upper South.[5]

The claim that hog cholera originated in the West is understandable. It was not accidental that the initial rapid growth and spread of hog cholera occurred in the region already noted for its concentration of hog raising and pork packing at Porkopolis as well as numerous other western river towns. The boom in hog population in this region during the first half of the nineteenth century brought unprecedented numbers of hogs in contact with each other, creating optimal conditions for the proliferation of disease and the magnification of its effects.

Regardless of origin, farmers had no solutions to the hog cholera problem. As the writer for the *Franklin Farmer* noted, "I, as a common sufferer with by brother farmers, have been trying to ascertain the cause of, and remedy for, the one which I have suffered the most by." Indeed, some of the symptoms he described are consistent with hog cholera, including purple discoloration of the body.[6] In 1860, the Kentucky legislature authorized a reward of a thousand dollars to "any person who may discover and make known the true cause of the disease called hog cholera, and a remedy that will cure the same." A few experts proposed cures. The "best and surest cure," offered by the Maryland State Agricultural Chemist,

**Figure 27**

The boom in hog raising in the nineteenth-century West, especially in the Ohio River Valley, provided ideal conditions for the spread of disease. Hogs kept in enclosures like this one during the fall fattening period could be easily fed from split-rail pens. Animals gathered from disparate farms for feeding in this way were susceptible to hog cholera and subsequently spread the disease everywhere they went. This photo dates to 1929, but it depicts a common feeding practice of the nineteenth century. Source: Ag Illustrated. Used with permission.

was a treatment of soda ash and barilla (alkali) that would reduce the inflammation in the lungs. According to the chemist, approximately 30 percent of pigs given the treatment lived. Of course, hog cholera rarely resulted in 100 percent mortality in a herd, making the validity of his claim dubious.[7]

The consensus of the 1850s was that hog cholera was preventable, even if there was no cure. This was consistent with prevailing medical views of the period, which attributed many diseases to environmental problems such as miasmas or vapors that emerged from swamps, cesspools, privy vaults, and dung heaps. Sanitarians believed that cleaning up these natural and man-made environmental problems would prevent disease. Dr. Edwin Snow of Rhode Island offered a sanitarian's perspective on hog cholera in an essay published in 1862. Snow urged farmers to remove the causes of disease. He conceded that atmospheric causes were hard to control, but reform-minded farmers could make animals stronger and less likely to succumb to hog cholera by cleaning up their barnyards and animals and providing proper feed. According to Snow, "the idea that hog cholera is contagious is abandoned, so far as I know, by all intelligent physicians at the present time."[8] Snow's anti-contagion perspective represented his best expert advice consistent with medical science, but farmers on the ground suspected otherwise, as seen in Robert Cartmell's 1859 testimony from Tennessee. His fear that the disease was "nearer me now than I like" suggests a grassroots perspective that hog cholera was, in fact, contagious.[9]

Hog cholera continued to plague hogs and hog farmers during the Civil War and for many years after. Reports from every region of the county indicated its presence, although it is impossible to know if all reports of the disease were actually incidents of hog cholera. While the Pacific Coast and the Northeast emerged comparatively unscathed, farmers in almost every other state endured significant outbreaks. James Mallory of Talladega County, Alabama, recorded in 1872 that most of his hogs died of hog cholera and that he failed to produce enough pork for family use for the first time in thirty years.[10]

During the 1870s, the problem escalated. In 1876, approximately four million American hogs, worth an estimated twenty million dollars, died of hog cholera. Illinois suffered almost one-fifth of the total loss in 1876, with Missouri, Iowa, and Indiana as the next most afflicted states, together accounting for almost half of the total loss. Florida, Alabama, Mississippi, and Louisiana also reported significant losses that year.[11] In an 1878 report to the National Agricultural Congress, veterinarian N. H. Paaren of Chicago stated that hog cholera had caused enormous losses in "the Middle and Western states." He stated that "almost entire stocks of individual farmers have been swept away in some counties of Illinois from the so-called hog cholera," with losses in 1877 numbering almost 359,000 hogs, exceeding the death toll of the previous year.[12]

In 1891, the Gebby family of Ohio watched helplessly as hog cholera moved through their herd. On September 26, Margaret Gebby noted that three animals were sick. The following day, she recorded, "the men doctored cholera hogs this morning," and repeated the procedure four days later, with Margaret conceding, "they are not any better." The first hog cholera death on the Gebby farm occurred on October 3. Two weeks later two more were dead. On October 22, they buried their eighth dead hog. By the end of November, at least ten mature hogs and a dozen young pigs were dead, some of which the Gebby family killed in an apparent attempt to prevent further suffering and the spread of the disease.[13]

Countless farmers attempted some sort of treatment on hogs, as either a preventative or a cure, claiming to have avoided significant losses. They attributed the cause of the disease to, among other things, bad feed, poor ventilation, and unsanitary conditions. Treatments included feeding charcoal, salt, red pepper, and copperas. Many of the reports were contradictory. In 1874, observers noted that hogs in apple orchards escaped the contagion in areas where hog cholera was present, while others claimed that hog cholera resulted from grazing hogs in an apple orchard. Some farmers and veterinarians claimed that hog cholera was due to excessive corn consumption, a common refrain in explaining the prevalent nature of hog cholera. One Iowa correspondent of the *Prairie Farmer* stated in 1872 that after losing several hogs to the disease, he orally administered up to half a pint of coal oil, which he alleged cured every case. Yet as one editor noted in 1875, "treatment of hog cholera is most unsatisfactory; in fact, all remedies are useless, when not administered as soon as the first symptoms appear." Hogs that survived these treatments might have survived anyway, since they may not have all been infected with hog cholera and, in fact, could have suffered from an entirely different malady.[14]

Some contemporaries suspected that this was the case. They believed that there were other diseases abroad, all labeled as hog cholera. An Ohio farmer suggested in 1872, "The name Hog Cholera may be applied to more than one virulent fatal disease among hogs." He was undoubtedly correct, which contributed to the confusion over why some hogs survived without treatment and why others survived despite treatments.[15]

## HOG CHOLERA AND THE USDA: SALMON, DORSET, AND HOG 844

The persistence of hog cholera, despite the numerous preventative measures and "cures," was a source of exasperation and fear, prompting Congress to respond in the 1870s and 1880s with funds for research and a new scientific organization to end the threat. In 1878, Congress appropriated funds to support a commission to investigate the disease. Nine scientists drafted a report synthesizing the collective ignorance on the issue, concluding that hog cholera was highly contagious and caused damage across the United States. The authors agreed that all proposed cures or preventatives were ineffective. In 1884, Congress authorized the creation of the Bureau of Animal Industry (BAI) within the US Department of Agriculture to conduct research on the many problems facing livestock producers, including hog cholera and Texas fever in cattle. Dr. Daniel E. Salmon, the first recipient of a doctor of veterinary medicine from an American institution, became head of the BAI.[16]

Salmon and his staff benefitted from the newly articulated germ theory of disease. Most American scientists quickly accepted Louis Pasteur's germ theory and agreed that the cause of hog cholera was likely a bacterium, although they disagreed about the strategy or techniques required to combat it. By 1885, Salmon claimed that there were two distinct diseases in the United States that had previously been labeled as hog cholera: the swine plague and hog cholera.[17]

One of Salmon's chief critics was Dr. Frank S. Billings, who not only challenged Salmon's scientific findings but also engaged in personal and public attacks on Salmon. Billings had hoped to obtain a position with the BAI in Nebraska to investigate hog cholera but, rebuffed by Salmon, instead was installed as director of the new Patho-Biological Laboratory at the University of Nebraska in 1886. In December 1887, he reported that he had identified the agent that caused Texas fever and was developing a vaccine for hog cholera. Billings attacked Salmon through the pages of *Nebraska Farmer*, where Billings was an associate editor. He refuted Salmon's claim that two distinct hog diseases were abroad, both labeled as hog cholera, and he further argued that the USDA was failing the American farmer.[18]

Under pressure to prove his claims about identifying the cause and cure of hog cholera through inoculation, Billings conducted public tests in 1888. The results were disappointing. The majority of the vaccinated hogs died. Billings blamed errors in the administration of the test and claimed that the study hogs had already contracted hog cholera prior to the test. Salmon requested that the USDA establish a commission that could clear him of Billings's charges and vindicate the work of the BAI. The 1889 commission report supported the BAI but, despite Salmon's cautions to the contrary, suggested that successful inoculation was likely and even imminent.[19]

Over the course of the 1890s, Salmon and the USDA failed to find a cure. Frustration over this failure and Salmon's pessimism gave credibility to Billings and others who accused the BAI of fraud and incompetence.

In 1892, Billings labeled the work of the BAI "a public scandal" in a self-published work. Billings charged that Salmon and the BAI falsified scientific studies, suppressed evidence, and engaged in "unmanly" attacks on Billings.[20] Salmon, besieged by Billings, was defensive. He believed in the validity of his work but recognized the scale and complexity of the problem. In remarks given in 1890, Salmon stated, "The prospect of discovering any kind of medicine which will prevent or cure the disease is not at all hopeful."[21]

Salmon's work at the BAI continued, despite the constant distractions of defending his integrity and that of the BAI from attacks by Billings. In the mid-1890s Salmon's principal researchers, Drs. Theobald Smith, F. L. Kilbourne, and Varanus A. Moore, developed a hog cholera serum. Researchers at the BAI injected hog cholera cultures into animals to determine whether they could acquire some form of immunity. If so, the researchers reasoned, their blood could be used to develop a serum to protect against hog cholera. In 1897 and 1898, the BAI conducted serum treatment fieldwork in Page County, Iowa, under the direction of Dr. Marion Dorset in 1897 and Dr. John McBirney in 1898. Results were inconclusive.

There was some evidence that BAI researchers were offtrack. As early as 1891, Smith noted that there were cases in which animals that exhibited hog cholera symptoms were not infected with the hog cholera bacillus. Furthermore, the disease produced by lab cultures was not especially contagious, and pigs infected with it were not immune to subsequent infections. Hog cholera, by contrast, was highly contagious, and the hogs that contracted the disease from other hogs were subsequently immune from future infection. In the Iowa fieldwork of 1897–1898, the BAI serum failed to improve mortality rates.[22]

The reason for these failures became apparent with the discovery of viruses. Researchers puzzled over the fact that they could pass bacteria-infected material through a fine-pore filter that would block the bacteria, yet the material still contained the infection. In 1903, BAI researchers Dr. Marion Dorset and Dr. Alexander de Schweinitz revealed that when they filtered the serum to separate out the hog cholera bacillus, the serum still produced hog cholera when injected into nonimmune hogs. True hog cholera, they concluded, was the product of a virus, not a bacillus. This was simultaneously a major advance in the understanding of the disease and a major blow to BAI credibility since Salmon had maintained since 1885 that a bacillus was the causative agent. Not surprisingly, critics attacked Dorset and de Schweinitz, but in the coming years researchers in the United States and Europe endorsed their work. Perhaps most importantly, Dr. Salmon publicly endorsed the research as legitimate in 1904, just one year after publication of their discovery.[23] Having successfully identified the cause, the quest for a vaccine or cure commenced. The twentieth century began with optimism that the threat could finally be managed.

Despite the discovery of the cause of hog cholera, it remained a persistent and devastating threat. In 1903 alone, losses due to the disease amounted to approximately fifty million dollars for American agriculture, causing hardship in every state. Between 1884 and 1913, the estimated loss to the American hog population due to hog cholera averaged 7.5 percent, with six years of losses over 10 percent. After the beginning of widespread immunization starting in 1915, the rate of losses declined, but still ranged between 3 and 5 percent prior to 1940. Not surprisingly, losses tended to be heaviest in the major hog-producing states. South Dakota farmer Ulbe Eringa endured three outbreaks of hog cholera on his farm between 1896 and 1926, including a 70 percent loss in 1906. Hog cholera devastated Boston's piggery on Deer Island in 1918.[24]

Change was coming, albeit too slowly for most farmers. This was due in large part to the work of Dr. Marion Dorset and hog number 844, a resident of the Bethesda, Maryland, Agricultural Experiment Station. Dorset hypothesized that he could create a hyperimmune hog through repeated injections of blood from diseased animals into a subject that had survived the disease. Starting in 1903, station hands injected 844 at multiple intervals with ever greater quantities of infected blood, up to 400 cc in one dose without displaying any symptoms. The hog was large, making blood draws difficult. Dr. William B. Niles suggested using a chisel and hammer to cut an inch off the end of 844's tail to facilitate the blood draw, which the technicians did to collect 100 cc of blood. The serum sample from the hyperimmune hog was subsequently injected into two susceptible pigs, one with some virus blood and the other without. Five weeks later, when those two pigs were injected with live hog cholera virus, the pig that received only serum died while the animal that received the serum-virus combination lived. This serum-virus treatment, commonly known as the simultaneous technique, was the long hoped for solution that would allow farmers to attack hog cholera. In a remarkable gesture, Dorset patented the technique and then donated the rights to the public.[25]

Extensive testing and discussion ensued along the Skunk River near Ames, Iowa, at the new Hog Cholera Research Station, established in 1905. A massive field study began in Iowa during the fall of 1907, with over two thousand animals tested with the serum-virus and serum. The results were overwhelmingly positive, confirming the utility of the simultaneous method in preventing hog cholera devastation.[26]

## HOG CHOLERA CONTROL

In 1908, a national conference of experiment station experts, BAI officials, and state veterinarians convened in Ames to discuss the next steps for research and the dissemination of vaccine. Delegates congratulated Dorset and Niles and urged cooperation between the states and the USDA to produce and distribute serum. Dorset's 1909 USDA bulletin was a summary of the research conducted to date, diagnostic techniques, and control measures. The editor of *Berkshire World and Corn Belt Stockman* subsequently claimed that hog cholera, "the old-time foe of hog-raising," no longer held the same power to terrorize farmers. "Science," he asserted, "has dealt it an effective blow," inaugurating a "new era in pork production."[27] By 1912, thirty states had either commenced serum manufacture or purchased serum from private manufacturers. During 1913 and 1914 the BAI organized a seventeen-county demonstration project in Georgia, Idaho, Indiana, Iowa, Kansas, Kentucky, Michigan, Minnesota, Missouri, Nebraska, Oklahoma, South Dakota, and Tennessee.[28] Tennessee's state veterinarian claimed in 1914 that the cost of the serum manufacturing plant had already been exceeded in the value of hogs saved by the use of state-manufactured serum.[29] It was, in fact, a new era in hog production. While hog cholera persisted, the prospect of control was conceivable.

Hog cholera control was demanding work. Harry S. Truman of Missouri helped a neighbor vaccinate his hogs in September 1912, describing the operation in a letter. "It was necessary to sneak up and grab a hind leg," he began, "then hold on until someone else got another hold wherever he could, and then proceed to throw Mr. Hog and sit on him until he got what the MO University says is good for him." Truman claimed that vaccinating hogs was excellent exercise, joking that it beat "Jack Johnson's whole training camp as a muscle toughener."[30]

Later that month, however, hog cholera devastated the Truman farm herd. Harry managed to market thirteen of his ninety hogs before they were symptomatic, but the remaining seventy-seven died. He masked his disappointment with humor, quipping that the hogs were within their rights to "kick the bucket," but he regretted that they did not die at the hands of the packers after he was paid. Given the timing of Truman's visit to his neighbor's farm to help vaccinate and the outbreak on his own property, it is possible that Truman brought the contagion to his own herd on his boots.[31]

Despite the efficacy of the serum-virus simultaneous technique, there were critics of the BAI and USDA. In 1913, veterinarian E. F. Lowry of Ottumwa, Iowa, attacked BAI scientists. He claimed that necropsies of infected hogs he conducted in the late 1890s revealed that all the infected animals were wormy. In a tract titled "Don't Vaccinate Your Hogs," Lowry concluded that worms, not germs, were responsible for the disease. Responsible farmers, he argued, needed to administer wormer. Not surprisingly, the last page of Lowry's tract was an advertisement for American Worm Expeller, manufactured by the American Livestock Powder Company of Shenandoah, Iowa.[32] The social landscape of hog cholera control remained characterized by misinformation.

More credible vaccination opponents included Dr. John Connaway of the University of Missouri. Connaway contended that the use of serum-virus was actually responsible for the spread of hog cholera. In his 1915 tract titled "Stamping Out Hog Cholera" published by *Missouri Farmer* magazine, Connaway argued that reckless vaccination by a "veterinary police force" and the administration of the serum were "doomed to failure." Farmer cooperation, quarantine, proper sanitation, control of carriers (sick hogs, dogs, streams, ponds, and barnyard fowl), parasite control, and good management were more important than vaccination. He endorsed vaccination on a limited basis, but only for farmers with infected herds, not as prophylaxis.[33] Connaway occupied a midpoint on the vaccination spectrum. He acknowledged its value but emphasized prevention through cultural practices. He was correct that the use of the serum-virus likely did reintroduce hog cholera in some places, but the use of the serum-virus lessened the severity of hog cholera losses. Furthermore, most farmers did

**Figure 28**

*Above left:* In this undated photo, men at the Hog Cholera Research Station in Ames, Iowa, prepare to extract blood from a hyperimmunized hog. Source: National Animal Disease Center, USDA. Used with permission.

**Figure 29**

*Above right:* After the tail was cut, one of the men collected blood in a pan. The infected blood from this hog was used to manufacture hog cholera serum. Source: National Animal Disease Center, USDA. Used with permission.

not immunize. In fact, a 1926 BAI estimate suggested that approximately 80 percent of American hogs were unvaccinated.

To reach these farmers, a host of companies manufactured and sold hog cholera "cures." The *Berkshire World and Corn Belt Stockman* dedicated several columns to debunking the claims of the promoters of a cure called "benetol" in 1914. The next year the Iowa Agricultural Experiment Station published the test results of a comparison of the BAI's Niles-Dorset serum with seven "so-called Hog Cholera Cures and Specifics." Not one of the tested products "protected hogs against attacks of hog cholera except [Niles-Dorset] hog cholera serum," developed by the BAI. In 1918, a con man hit Linn County, Iowa, selling a medicine sure to prevent hog cholera if administered every week. Fakes and low-quality serum not only failed to prevent the disease but also convinced some farmers that serum treatment was ineffective.[34] As one veterinarian claimed, "Hog raisers have spent hundreds of thousands of dollars in purchasing this worthless material, and, as a result, the farmer has formed the opinion that practically all hog-cholera remedies are fakes" and were, consequently, unwilling to use the Niles-Dorset serum even when provided free of charge.[35]

Quackery aside, most farm experts emphasized the importance of farm sanitation as part of the hog-cholera prevention message. Rotation of hog yards, with old lots plowed and sown to pasture, was a technique to prevent carryover from one year to the next. Farmers could build concrete hog wallows filled with water and disinfectant. They could disinfect hog houses and sheds with carbolic acid solution. Even sanitizing footwear

**Figure 30**

Two employees at the Hog Cholera Research Station administer live-virus serum to a hog. Source: National Animal Disease Center, USDA. Used with permission.

before leaving a farm reduced the risk of carrying the disease from one farm to another.[36]

There was also new hope for an improved hog cholera vaccine. In the early 1930s, the BAI resumed vaccine research. The delay in research was probably due to the success of the serum-virus immunization program in limiting the worst outbreaks of the disease. The chief of the BAI announced in 1934 that studies had been conducted on producing an attenuated virus, a less potent strain that reduced the risk of new infections with the serum-virus treatment. Tests conducted in the late 1930s were promising, although vaccinated animals required up to three weeks for immunity to develop and that immunity did not last as long in vaccinated animals as it did in serum-virus treated hogs. Hopes for the development of an improved vaccine were limited. The Great Depression occupied the attention of many Americans. Furthermore, the outbreak of war in Asia and Europe precluded a large-scale effort to improve and disseminate vaccines.[37]

Thanks to the widespread use of serum to limit the spread of hog cholera, by the 1940s hogs lived healthier lives than their late nineteenth-century forebears. Farmers who immunized had a protection rate of over 90 percent, which was an excellent result since an infected herd could sustain a complete loss. Death losses due to hog cholera were at or below 3 percent per year throughout the 1930s. While still significant, this represented a major decline from the staggering losses of previous decades. One BAI veterinary inspector stated in 1935 that the "diminution" of hog cholera resulted in more attention paid to other diseases. Specifically, he listed those that "in time of unusual prevalence of hog cholera, go unnoticed or possibly diagnosed as cholera itself," including "anthrax, epilepsy, gastroenteritis, necrobacillosis, pleurisy, pneumonia, poisoning, swine plague (hemorrhagic septicemia), swine erysipelas, tuberculosis, and worms."[38]

Encouraged by progress in vaccine research, in 1950 representatives of the American Veterinary Medical Association and the US Livestock Sanitary Association urged the creation of a government committee to develop a hog cholera eradication program in the United States. This group advocated for the replacement of the virulent virus with a modified vaccine accompanied by an aggressive public education campaign to alert farmers and consumers about the benefits of eradication.[39]

In 1961, the secretary of agriculture issued a report on the prospects of hog cholera eradication, highlighting the fact that hog cholera, even at the relatively low infection rates of the 1940s and 1950s, was responsible for a loss of approximately forty to sixty million dollars every year, although the actual figure was likely higher because most states did not have a system for reporting outbreaks. Regardless of the financial loss, those numbers obscure the fact that millions of animals continued to suffer and die from the disease. Inspired by the report, Congress passed a bill to create a national hog cholera eradication program, which President John F. Kennedy signed into law in September 1961.[40]

The eradication effort was organized around what was labeled the Nine-Point Program. The nine actions included the elimination of the virulent hog cholera virus as part of the current inoculation effort, a prohibition on feeding raw garbage to swine, the reporting of outbreaks, quarantines of infected farms, increased vaccination rates, the tracking of hog movements, the cleaning and disinfecting of infected farms and vehicles, more research, and a public education campaign. None of these actions would have surprised anyone who knew anything about hog cholera or its history. The difficulty, however, was coordination and implementation.[41]

Members of the USDA and the US Animal Health Association met and developed a four-phase implementation program in late 1962. The first phase, "Preparation," would be organization, in which state and county committees would establish procedures for reporting suspected outbreaks, investigate suspected outbreaks, and develop a standard diagnostic procedure. The goal was to have all states participating in this program by the end of 1964.[42] The next phase, "Reduction of Incidence," would take place from 1965 through 1967 and involved quarantines of suspected herds, restricted movement from quarantined herds, inspection of all marketed swine that would return to farms, and the development of records for all swine dealers. Phase 3, "Elimination of Outbreaks," was to occur from 1968 through 1969 with an emphasis on the elimination of infected herds and proper disposal of all infected animals and carcasses. Owners of condemned animals received indemnification from state or federal agencies. All infected farms and swine-handling facilities were to be cleaned and disinfected. The final phase of the campaign was "Protection against Reinfection." In this phase states would have no diagnosed cases, the use of the live virus vaccine would cease, all inactivated virus would be reported to state authorities, and each state would impose a twenty-one-day isolation period for all swine imported into the state. The only exception would be for hogs arriving from hog-cholera-free states. The target for every state to be free of hog cholera was the end of 1972.

Progress was slower than expected. There were only 118 recorded cases in 1971, with twenty-nine states reporting no infections in 1971. In a 1972 report, the USDA indicated that only thirty-two states were free of hog cholera but that the incidence of the disease was the lowest on record.[43] As one hog cholera expert reflected, the goals were reasonable but variability in state and federal funding coupled with complacency on the part of farmers meant that it was 1977 before the USDA declared every state "hog cholera free."[44] The Pacific Northwest and western states were the first areas to reach that status. By 1974, only Texas, North Carolina, New Jersey, and Puerto Rico remained on the list.[45] In 1978, Secretary of Agriculture Bob Bergland pronounced that the United States was hog cholera free. It had been a difficult and long process. In the early twentieth century, the dream was control, not eradication. Once the infection rate was low enough, however, scientists and policy makers began to envision an agricultural system freed from one of its largest threats. Within twenty years, that vision was reality.

## NEW DISEASES FILL THE VOID

New diseases, however, rose to significance in the wake of hog cholera eradication, posing challenges for an increasingly centralized and intensively managed hog industry. By enabling the development of large-scale hog operations, disease control experts in effect created the conditions for new diseases to thrive. Hog farmers began to grapple with the spread of transmissible gastroenteritis (TGE), a virus that first appeared in 1946 but emerged as a major problem only in the 1980s. TGE causes diarrhea and vomiting, leading to near 100 percent mortality in unweaned pigs two weeks of age and younger. One observer in the hog barns in the 1990s, anticipating a look at the cute, young pigs, was horrified to see the floor "covered with tiny dead pigs," and the few live ones sprawled on all fours, shaking "with tremors, and soaking wet in their own milky diarrhea and vomit" due to TGE.[46]

Numerous other pathogens stalk the ever-larger American hog farms, including porcine reproductive and respiratory syndrome (PRRS). The very growth and concentration in the industry magnified the impact of these diseases, just as it had for hog cholera in the mid-nineteenth century. In 2007, one journalist explained, "hog density creates PRRS health challenges," including the opportunity for repeated infection of new strains of the disease.[47] Biosecurity protocols are in place on operations of all sizes to minimize the risk of contagion. Procedures typically consist of washing trucks, changing clothes and shoes on entering and exiting the facility, showering, and requiring visitors and employees to observe minimum downtime periods between visits to other hog facilities, to name just a few.[48] In some cases, biosecurity procedures are so strict that companies restrict after-hours contact for employees responsible for different stages of hog production.[49]

Most recently, the industry experienced a new killer, porcine epidemic diarrhea virus (PEDv), part of a family of viruses labeled swine enteric coronavirus diseases (SECD). While scientists identified PEDv in the 1970s, it appeared in the United States only in 2013. By July of that year, it had spread to fourteen states and by mid-2014 had resulted in an estimated loss of 5 to 10 percent of the entire US hog population.[50] During the first year of the outbreak, approximately eight million young American pigs died of PEDv.[51] As one Iowa farmer reported from his farm, "It was about 850 pigs didn't make it. For three weeks it was 100 percent death."[52] Alfred Smith, owner of Garland Farms near Plains, Kansas, stated, "We are out in the middle of nowhere but that didn't matter. We lost four weeks' worth of pigs. It's similar to TGE, but worse."[53] Again, there was no cure. Even good management practice is not a guarantee against infection. Discussing PRRS in 2004, a Minnesota veterinarian stated, "We build fences, truck washes, and it doesn't stop it. We can't say to producers, 'If you do this, you won't get it.' We can clean up a PRRS farm, but the chances of it staying clean are very slim."[54]

The rapid emergence and spread of PEDv, PRRS, and other diseases such as porcine delta coronavirus (PDCoV) are constant reminders of the risks of large-scale hog farming. Hog cholera emerged in the West during the mid-nineteenth century as farmers there engaged in hog farming at an unprecedented scale. It continued to ravage American herds everywhere, but especially in the South and Midwest because those regions were the nation's greatest hog producers. Yet, having justifiably eliminated hog cholera at considerable expense, society must reckon with the fact that there is no end to disease risks associated with animal agriculture, especially the kind that relies on massive scale. Size has magnified the risks, despite the best efforts to keep pigs alive.

Through it all, the US government has been present, allied with agriculture and the meat industry. In June 2014, the USDA issued a license for the production of an experimental vaccine to control PEDv, a move that followed the release of $26.2 million to fund control efforts. As the press release from the Veterinary Services office of the USDA's Animal and Plant Health Inspection Service noted, the vaccine license was a step to "help the pork industry and producers." The attempt was just the latest in a long history of government intervention to shore up the swine industry in transcending traditional limits on the scale of production.

CHAPTER 7

# MAKING BACON AND WHITE MEAT

At the middle of the nineteenth century, Americans favored beef over pork, despite Dr. John Wilson's assertion in 1860 that the United States was a great "Hog-eating Confederacy." While Americans everywhere relied on pork as an important part of their diet, it was the southerners and westerners Wilson knew best who depended on swine the most. Pork displaced the English preference for mutton by the nineteenth century, but it still trailed beef. Only with the onset of the Great Depression and the spread of poverty did pork challenge beef for primacy. The post–World War II period, however, saw a drastic reversal for both beef and pork. Chicken, a status meat like beef, became America's favored meat choice in the late twentieth century thanks to consumer desire for lean meat and the meat industry's ability to manipulate species and markets. Even so, salty preserved pork gained new adherents by the twenty-first century. Foodies and gourmands rediscovered old-fashioned pork, charcuterie, and bacon, sparking new interest in the fattier cuts that consumers had rejected presumably in their concern about dietary fat.

## PORK AND POST–CIVIL WAR RURAL AMERICA

To understand the persistence of America's beef preference and the rise of pork, it is useful to start in the South, the heart of Wilson's "Hog-eating Confederacy." The defeat of the Confederate States of America in the Civil War enshrined the place of pork in the American South, largely due to the persistence of an impoverished rural working class. Historian Joe Gray Taylor argued that southerners ate more pork after the war than they did before due to the rise of sharecropping. For plantation owners after Emancipation, land was most valuable for nonperishable commodity production, not food production. In the days of slavery, plantation owners assumed the costs of providing food, shelter, clothing, and medical care for enslaved workers. In this reorganized social and economic space, landowners often set up commissaries or stores on their farm where sharecroppers and tenants purchased food and supplies, most often salt pork or bacon, cornmeal, and sorghum molasses: all commodities that were easy to store and transport and were already common fare for poor southerners. They sold pork, cornmeal, and molasses at significant markups, profiting from the worker's dietary deficit. Plantation owners turned what had been an expense of provisioning slaves into income. It is no exaggeration that the boom in southern cotton commodity production after the Civil War depended on the lowly American hog as well as sharecroppers and tenants.

Fried pork and grease was the rule for these southerners. At Waverly Plantation, located near Columbus, Mississippi, farm tenants ate mostly pork, although rabbit, squirrel, opossum, deer, fish, mussels, waterfowl, and even other domestic animals such as goat, sheep, and cattle were all part of the diet, albeit in small quantities. The predominance of pork there was due to the fact that most of the meat sold at the commissary was pork.[1] An 1890s study of the foodways practiced by rural African Americans in Alabama explained that "the salt pork is sliced thin and fried until very brown and much of the grease tried out. Molasses from cane or sorghum is added to the fat, making what is known as 'sap,' which is eaten with the corn bread. Hot water sweetened with molasses is used as a beverage." A corn bread recipe called "crackling bread" was prepared by cooking bacon until brittle, then crushing and mixing it with cornmeal, water, soda, and salt and baking the dough. This was common fare in the "cabins on the plantations of the 'black belt,' three times a day."[2] It was, above all, southern country food.

In 1875, traveler Edward King noted that while southern cities boasted highbrow cuisine, one did not have to travel far from the city to find that "one comes into a region where coarse bread, coarser pork, and few stunted vegetables are the only articles of diet upon the farmers' tables." The meat was "fried and greasy ham, or bacon, as it us usually called in the South." King complained that on the rare occasion that he encountered beefsteak, it was "remorselessly fried until not a particle of juice remained."[3]

Surging demand for pork due to the transition to tenancy and share-cropping as well as wartime decimation of southern herds compelled southerners to reestablish their dependency on midwestern pork. One Virginian lamented in 1869 that the South was doomed to poverty as long as they relied on the Northeast for manufactured goods and the West for horses, mules, and bacon. The closing of the range, discussed elsewhere, exacerbated the food dependency problem because the tenants and sharecroppers of the South lost the ability to raise their own meat and became pork and bacon buyers rather than hog raisers.[4]

Rural people in other places and regions, from lumber camps and gold fields to cattle trails and farms, continued to rely on pork as a significant part of their diet in the post–Civil War years. One cowboy recalled that rations on the trail to Kansas during the 1870s consisted of "a sack of Navy beans, a sack of dried apples, coffee and a side of [fat] bacon."[5] Swedish immigrant Ida Lindgren remarked on the regularity of boiled and fried pork in the diet on her farm near Manhattan, Kansas, in 1870. When Lindgren arrived at Lake Sibley, she noted that for two weeks they ate "fried pork and potatoes, butter, bread, and coffee for every meal, *every meal*!"[6] Francis Newlands of Nevada (and future US senator) complained in the 1880s that there were plenty of cattle in Nevada, but "the great reliance is ham and bacon swimming in fat."[7] An 1896 dietary study of Missouri farmers indicated that they ate 56.9 percent of their meat as pork, followed by 20.9 percent as beef and 12.9 percent as poultry. As the author of the study explained, "The farmer does not have easy access to the butcher's shops and furthermore has no conveniences for keeping fresh beef," leaving pork as the practical, everyday farm food. Given the ubiquity of hogs on farms across the country, the reliance on pork was a common feature of rural life.[8]

Annual fall butchering was a common practice, with farm diarists noting carcass weight and the division of family labor. Iowans John and Tacy Savage butchered one or more hogs for home use every year throughout the 1860s and 1870s, occasionally selling some fresh or cured pork. The Gebby family of Logan County, Ohio, raised hogs for the market but also for home use in the 1880s and 1890s. The family butchered two hogs in 1895 and 1896, one in 1888, and three in 1889. They must have also butchered in 1894, since Margaret Gebby reported hanging meat in the smokehouse that year. The family made much of that meat into sausage, with a low of forty-eight pounds and a high of seventy-five pounds recorded in Margaret's diary. In February 1897, Isaac Carr of Iowa reported making his brine for curing ham, which consisted of "8 or 9 pounds of sugar, salt peter, and salt until it holds up an egg." Farmers who lived near town could even take their livestock to town for butchering, like the Kimball family of Jones County, Iowa.[9] Reliance on the pork barrel and smokehouse was common in every farm community.

## PORK AND LARD IN INDUSTRIAL AMERICA

While pork was king of meat for rural Americans, the old hierarchy of beef, mutton, and pork of the prewar years lasted longer in towns and cities. Historian Harvey Levenstein claimed that the urban middle class developed a "beef and potatoes syndrome" during the late nineteenth century, with steak valued as the perfect meat. According to an 1882 cookbook, beef was "Bible and chemically sanctioned, purposely designed for man." Dietary studies of Missouri businessmen revealed that they ate 49.6 percent of their meat as beef and 23.5 percent as pork, while Missouri professionals such as doctors and lawyers consumed 47 percent of their meat as beef and 25.3 percent as pork. Late nineteenth-century steak was, as historian Richard Hooker stated, "increasingly defined by specific cuts: sirloin steak, porterhouse steak, tenderloin steak, and filet of beef," often sliced thin, fried, and served with onion, mushroom, or tomato.[10]

For members of the urban middle and upper classes, pork was inferior. It was too "coarse" for the new professionals, in part because it was too fat for the new sedentary lifestyle. The author of the *Housekeeper's Encyclopedia*, published in 1864, noted that previous generations did extreme manual labor "with no aid from machinery." Using only hand tools, they chopped wood, mowed and raked hay, harvested and threshed grain, and walked behind a plow for miles. For these people, the author claimed, "pork was needed to sustain them."[11]

Bolling Hubbard of Virginia affirmed the association of pork with country people when he moved to New York City in 1871. The upwardly aspiring Hubbard lamented that his father-in-law refused to provide financial assistance, complaining that his wife must be reduced to wearing calico and eating "hog and hominy," two characteristics of rural southerners that he hoped to avoid.[12] A writer for *Good Housekeeping* went so far in 1890 as to claim that "as an article of food, pork, of late years, does not generally meet the approval of intelligent people."[13] When novelist Theodore Dreiser depicted the gulf between urban success and poverty in *Sister Carrie* (1900), his protagonist was excited and intimidated by the $1.25 charge for sirloin steak and mushrooms featured on a Chicago restaurant menu.

Yet pork was a common and reliable food for urban workers. Historian Fred Shannon suggested that unskilled urban laborers who earned $500 or less a year (approximately $1.50 a day) could afford to purchase a "pound of beefsteak at a cost of 15 cents" on Sundays. But on most days, their fare consisted of inexpensive and monotonous salt pork and beans.[14] More commonly, it was pork in the bean pot on Saturday night, at least for New Englanders. According to one 1877 recipe, baked beans consisted of one and a half pints of navy beans and a quarter pound of pickled pork.[15] Skilled native-born workers ate less pork than unskilled workers or immigrants. In turn-of-the-century Missouri, skilled workers ate 40.2 percent of their meat as beef and 32.4 percent as pork.[16] Urban immigrants to cities brought their Old World customs, and for the new wave of German immigrants of the late nineteenth century, that meant regular servings of pork, including pig's feet and headcheese. When immigrant Christian Kirst arrived in Pittsburgh in 1881, his "circle of friends" brought numerous food items for him, notably lard and bacon.[17]

Scrapple took on renewed significance for urban working-class America where there had been significant German settlement. This dish, consisting of scraps and trimmings from the boiled hog's head mixed with cornmeal and seasoning, was popular on Pennsylvania and Delaware farms in the eighteenth and early nineteenth centuries. Later, when those Mid-Atlantic cities grew, scrapple moved to the city, where it earned a reputation for being a suitable food for the poor. In Philadelphia, "distributing the scrapple" actually became a euphemism for dispensing political favors and patronage. More prosperous scrapple promoters claimed that it was a welcome change for those "who have become wearied of the eternal round of steak, [and] chop." Detractors noted that "a quart of sawdust and a pound of tallow dips, and with wicks included" would make a better dish.[18] Still, a machinist who earned approximately $100 per month in New York City spent $1.50 per day on food in 1872: 45 cents for breakfast, 70 cents for midday dinner, and 35 cents for supper. "Could I do it for less eating scrapple?" he asked. Yes, but it would not be favored "every day in the week," and it was not as cheap "as pork and beans."[19]

Urban African Americans consumed more pork than beef. A dietary study of poor African Americans in Philadelphia and Washington, DC, indicated that pork sausage was a primary food, bacon was of secondary importance, and ham was peripheral. This was in contrast to whites of the same economic group, for whom sausage was of only secondary importance. Bones recovered from two Annapolis, Maryland, African American residential properties of the post-1874 period indicated that those residents consumed much more pork and fish than beef. The similarity to other urban and rural southerners throughout the nineteenth century and free African Americans before the Civil War suggests that more than economics was at work. Food selection for these families may have affirmed cultural identity and distinctiveness as much as it reflected economic status.[20]

For all Americans, regardless of status, location, and ethnicity, lard was critical in the kitchen. It was especially critical to New Mexico's Hispanic population. The diet of Hispanics in the Las Cruces area, the subject of an 1895 study, was heavy in vegetables and light in meat, with beans and lentils composing much of the protein. Hispanic families, however, purchased and used significant quantities of lard and tallow as the principal sources of fat in their diet. Farmers often recorded the amount of lard rendered at butchering time. In 1877, Sarah Jane Kimball of Iowa recorded preparing sausage and headcheese as well as rendering lard at butchering time, just like thousands of other families. Margaret Gebby processed fifteen gallons of lard from two hogs in 1889 and fourteen to fifteen gallons from two hogs in 1896 on her Ohio farm. Lard was an important retail product, too, for those who lived in towns and cities.[21]

In the 1880s, a lard adulteration scandal led some consumers to question the healthiness of lard as a food product. The expansion of cotton production after the Civil War, thanks to the rise of sharecropping and tenancy, led ginners and processors to develop industrial application for oil-rich cottonseeds, most often ground up or pressed for the oil, which made a rich livestock feed supplement. It was also a suitable cooking shortening and, when mixed with lard, was unrecognizable as vegetable oil. In their attempt to cut costs, pork processors added inexpensive vegetable oils such as cotton seed oil. The packers, however, did not alert consumers to this practice, leading to a public scandal.

There was an irony in the adulterated lard controversy. Many Americans viewed so-called pure lard as inferior to the very product used in the adulteration. Marketing by cottonseed oil processors in the 1890s portrayed lard as an undesirable and unhealthy product. Cottolene, a lard substitute developed in the late 1860s by N. K. Fairbank Company of Chicago, ultimately rivaled lard in the kitchen. Fairbank advertisements of the 1890s praised the purity of vegetable oil and attacked lard as unclean, unwholesome, and indigestible. They claimed that lard was simply grease and that food prepared with it was rich, soggy, and greasy. Advertisers asserted that lard possessed a "hoggish smell" and that food fried with Cottolene was better, even healthier than that fried with lard. Those prone to gastrointestinal complaints such as "children and dyspeptics" could consume Cottolene "with the utmost enjoyment and no fear of danger," while lard caused dyspepsia (indigestion). Cottolene, they contended, possessed curative powers. "If the Doctor Did Your Cooking," one advertisement read, "there would never be an ounce of lard used in your kitchen." Many consumers likely ignored this extreme rhetoric, but advertisers repeated the claims for the superiority of vegetable shortening over lard well into the twentieth century. Lard stigma became mainstream.[22]

Even so, lard and pork remained omnipresent in the American diet, with pork rivaling beef in the first half of the twentieth century. Many consumers preferred beef over pork, but the reality was that pork remained a low-cost meat. As a writer for *Wallaces' Farmer and Iowa Homestead* stated in 1907, "our friend the hog" fed industrial and farm laborers and kept skillets greased around the world, serving as a bulwark against hunger and privation for the global laboring population.[23] In 1908, beef consumption constituted approximately 48.6 percent of the American red-meat diet; pork, 47.5 percent. By 1940, however, pork overtook beef with 51.6 percent and beef at 43.8 percent. Wartime prosperity brought beef to the forefront again by 1943. Such figures obscure regional and class variations, but they do indicate the continued popularity of pork in a largely working-class nation. Steak remained high-status food, and most pork reflected a lower status. Averages also obscure the changes in the kinds of pork that Americans ate.

Preferences and tastes were changing. As historian Roger Horowitz noted, by the mid-twentieth century processors changed pork to meet the times.[24] The old-fashioned fatty barrel pork of the nineteenth century faded in importance. Artificial refrigeration made it easier to preserve and market fresh pork. Ships' crews had relied on barrel and salt pork for generations, but the introduction of onboard cold storage meant that sailors could consume a greater variety of foods, not to mention more fresh meat and less salty "junk."[25]

Demand for bacon and ham was in decline too, but packers found ways to maintain the viability of these products. Both bacon and ham involved time-consuming and costly curing techniques that lasted up to three months, so packers developed wet cures to speed the process. They added borax to the sugar, salt, and saltpeter brine to prevent the growth of unwanted bacteria and injected the brine into the meat. Although the USDA banned the use of borax in 1906, by the mid-1920s packers had substituted sodium nitrite to the wet cure and pumped it through the veins of hams. The new ham, however, was softer, less flavorful, and more watery than old-fashioned dry-cured hams. Urban consumers accepted this transformation, however, in part because new ham was less expensive than the old.[26]

## PERSISTENCE AND NEW THREATS: THE SOUTH, IMMIGRANTS, TRICHINA, AND FAT

In the midst of the urbanization and industrialization of the United States, pork remained important, especially in the South. In 1932, sociologist Rupert Vance claimed the typical meal of southern sharecroppers, white and black, was corn bread and pork. While pork was often fried, most often cooks boiled it with vegetables and greens in season. A mid-1920s study of rural Georgians (including farm owners, tenants, and sharecroppers) indicated that they consumed almost four times more pork than beef. African American sharecroppers who lived in Leon County, Florida, in the early 1900s purchased most of their pork and lard. Salt pork priced at ten cents per pound constituted a major portion of their pork purchases.[27]

Even among rural southerners, however, there was diversity in the role of hogs on farms and pork in the diet. Fatback and sowbelly remained as food for poor people. Sociologist Arthur Raper studied the social structure of Georgia's Macon and Greene Counties during the 1920s and 1930s and concluded that landless farmers ate fried fat pork for every meal, served with corn bread and molasses. Historian Frederick Douglass Opie asserted that poor people most commonly consumed "fried pork, fried fat pork, fried pork shoulder, and more fried side meat."[28] Rural southerners, white and black, prepared the intestines, known as chitterlings. In his memoir of growing up in the 1920s and 1930s, Jimmye Hillman of Greene County, Mississippi, recalled that the dish was a favorite of his father and the family's African American cook. Hillman assisted by turning the intestines inside out and scrubbing them with a brush and vinegar water, a process known as "ridding the guts." They then fried and served the intestines with Tabasco or Worcestershire sauce or sometimes boiled them, seasoned with salt, pepper, and onion, served alongside beets or turnips.[29] Poor rural African Americans often enjoyed distinctive dishes that their white counterparts did not prepare, including pork stew made with inexpensive parts of the carcass such as ears, feet, and backbone. In general, though, farm families that kept and slaughtered their own pigs ate more of the carcass than their small-town counterparts did. Hillman estimated that for his family pork accounted for approximately three-fourths of the meat supply and all of the cooking fat. Other meats on the Hillman table included beef, mutton, and game, but nothing compared to pork on their farm.

Southern textile mill workers ate purchased pork every day and consequently had less variety in cuts than farmers. Bacon and salt pork were the most commonly served meats on the tables of mill workers. Southerners sometimes referred to unnamed "meat" on their tables, but

**Figure 31**

Lard was a valuable and ubiquitous part of the American diet before World War II. In 1939, this artist prepared a lard sculpture at the Indiana State Fair touting its benefits as an economical cooking product. Source: Ag Illustrated. Used with permission.

as historian Joe Gray Taylor claimed, in the South "meat and pork were synonyms, and it would almost be accurate to say that meat and bacon were synonyms; if a southerner meant ham, shoulder, or jowls, he would say so."[30]

African Americans who moved north as part of the Great Migration brought rural pork culture with them to the city. Between 1910 and the 1960s, at least six million people left the South in search of jobs and education and to escape the worst features of Jim Crow oppression. During this period, barbeque restaurants became fixtures in almost every northern city. Two of Kansas City's legendary restaurants, Arthur Bryant's and Gates BBQ, traced their origins to Henry Perry, "the Barbeque King," who opened shop in Kansas City in 1907.[31] Pig's feet, ham hocks, and chitterlings, all associated with eating "low on the hog," were part of the diaspora. In Harlem, these foods were staples. One Harlem food vendor, Mary Dean, was locally renowned as Pig Foot Mary due to her specialty. In 1928, a writer asserted that pork was "the leading article of flesh in the diet" in Harlem. A reporter there remarked that Harlem residents regularly enjoyed greens "cooked with some sort of pork."[32]

Outside of African American urban enclaves, rural people and immigrants continued to be the northerners who ate the most pork. A 1942 study of food practices in rural downstate Illinois indicated that fat pork was the standard breakfast meat for breakfast and boiled or fried pork was the staple for noon dinner.[33] On the Hamilton farm in Iowa during the 1920s and 1930s, sausage and bacon produced on the farm were winter and spring breakfast staples. They used lard every day. Hamilton's mother rubbed a piece of ham rind or fat pork over the hot griddle to keep pancakes from sticking. In the summer when the pork was gone, the Hamiltons purchased beef or butchered their older chickens for the table. At butchering time, they preserved the prized loin by cooking it, placing the slices in a jar, and covering the whole with melted lard as a sealant.[34]

Pork retained a close association with European immigrants. In a 1921 discussion of immigrants in the Midwest, one observer noted that American-born country school teachers (mostly young, unmarried women) complained about working in immigrant communities. It was common for teachers to room with pupils' families, which meant that in immigrant communities the fare frequently included "pork and sauerkraut, sour milk, herring, onions, etc." American-born educators viewed these foods as ethnic and therefore less desirable. In the language of the resurgent Ku Klux Klan, those who consumed them were not fully "100% American."[35]

The stigmatization of pork was not only due to its association with immigrants but also due to an increase in trichinosis infections. The trichina threat was on the rise during the first half of the twentieth century, affecting approximately 16 percent of the US population by the 1940s. The overwhelming majority of those cases were asymptomatic, but for those few thousand people who reported symptoms, it was a painful and prolonged experience that occasionally resulted in death. Trichinosis caused approximately twenty deaths per year during this period.[36]

The trichinosis problem grew with the popularity of garbage feeding. Slaughterhouse offal and garbage included raw or undercooked pork scraps that often contained trichina cysts. Hogs that consumed infested waste spread the worm. By 1929, the USDA concluded that uncooked garbage was the single greatest culprit in spreading trichina.[37] In 1941, the New York State Trichinosis Commission reported that pigs fed on slaughterhouse offal were the most common source of infection because people consumed trichina-infected pork chops, fresh ham, pork loins, sausages, and other pork products. The commissioners warned New Yorkers to cook pork thoroughly and urged lawmakers to require garbage feeders to cook market refuse to kill the parasite.[38]

New concerns about diets further stigmatized pork. Government officials emphasized the importance of consuming lean meats to combat obesity, America's new nemesis. In 1923, they cautioned Americans against too much "bacon, salt pork, fat pork sausage, and cream" because those foods were judged "too fat" in relation to the amount of protein they possessed. Pork was conspicuous by its absence in a set of proposed menus for a week of balanced eating. The government menus included bacon for breakfast only two days a week and one supper of boiled potatoes served with gravy made with bacon fat.[39]

Government warnings about excessive fat consumption notwithstanding, the packing industry rehabilitated bacon, one of the fattiest cuts, for middle-class urban consumers. Packers developed sliced bacon in the mid-1910s as a convenience food to sustain its consumption. Earlier generations of middle-class Americans in northern cities relied on young Irish or Scandinavian immigrant women to work for low wages doing housework and cooking, including slicing bacon for the skillet. By the 1920s, household help was on the decline. Federal policy constricted the flow of immigrants in the name of "100% Americanism" at the same time as young women abandoned domestic service to pursue jobs with more independence and better wages such as office and clerical work or, in some cases, to become the next "It Girl" in silent films. Skillet-ready sliced bacon improved bacon's appeal for middle-class women who could not or would not pay for scarce and expensive household help. By 1941, transparent cellophane packaging enabled the consumer (invariably depicted as a female homemaker) to "see just what she is getting, safely protected from dirt and dust," with the brand name on the label to assure her "that she can get exactly the same quality every time."[40] Seeing the product gave consumers the confidence that there was actually a streak of lean in the bacon.

One of the most celebrated pork products in the American diet was Spam. Its name was a blend of the words "spicy" and "ham," suggesting both a distinctive flavor and the presence of ham, the top-valued pork

**Figure 32**
This photograph from the late 1930s shows a woman making chitterlings, or chitlins, on a farm near Maxton, North Carolina. The use of chitlins by poor rural southerners indicated the importance of eating all parts of the hog. Source: Library of Congress. Public domain.

cut. A creation of Jay Hormel, Spam consisted of ham and pork pieces, in particular pork shoulder that was otherwise destined for sausage. Combined with sugar, salt, potato starch, and sodium nitrite to maintain a pink color, Spam was unlike other processed pork products. Vacuum-sealed at the factory, Spam required no refrigeration and would last indefinitely in the pantry. Other packers introduced versions of precooked, tinned pork: Armour made Treet, Wilson and Company gave consumers Mor, and Swift produced Prem (short for Premium ham), but Spam dominated the marketplace. Consumer surveys indicated that by 1940 approximately 70 percent of American households used Spam, outpacing competitors by a two-to-one margin.[41]

Consumption of Spam and other tinned pork soared thanks to the Great Depression and World War II. Spam was convenient, defied seasonality, maximized convenience, and was inexpensive compared to other meats. It was precooked and therefore had a long shelf life, was high in protein and salt, and was easily transported. During the war, Spam and its imitators were exempt from rationing, which made it a good protein source for civilians who endured reduced meat options. Advertisers emphasized Spam's ease of preparation for families in which both women and men were out of the home doing war work. For the majority of GIs, however, most of the tinned meat they consumed was not Spam, although

**Figure 33**
African Americans brought southern foodways to northern cities. This 1942 photograph shows a grocery in an African American neighborhood of Detroit. The promotions on the shop windows featured many parts of the hog carcass that would later become known as soul food: chitlins and hog maws, liver and lights (lungs), hog fries (testicles), tongue, kidneys, and lard. Source: Library of Congress. Public domain.

it occasionally arrived on the front lines as an authorized substitute for luncheon meat. Service personnel ate tinned "luncheon meat" prepared to government specifications. Nevertheless, Spam was part of the B Ration, served primarily to troops stationed in the United States and as part of Red Cross shipments overseas. Great Britain and the Soviet Union received the most wartime Spam exports, affirming the place of American pork on tables around the world.[42]

## PORK ECLIPSED: CHICKEN AND THE OTHER WHITE MEAT

After surging during the 1930s due to the relatively low cost of pork compared to other meat, the position of pork in the American diet was more important than it had been for decades. By 1942, farmers, especially middling farmers, consumed more pork per capita than the rural nonfarm and urban population. The only exception was that of the poorest city dwellers, whose pork consumption rivaled that of farmers.[43] Pork reached an all-time high in 1944, due to its low cost. Just one year later, however, American annual per capita consumption of beef in the nation was 71.3 pounds, followed by 66.6 pounds of pork and 25 pounds of poultry. Pork continued to slide in the postwar period. By the 1980s, pork consumption stabilized in third position in the American meat hierarchy behind chicken and beef.[44]

The two most salient characteristics of postwar meat consumption were the return of beef as America's top meat in the short term and the ascendance of chicken in the long term. For the nation's growing middle class, broiled beef or grilled steak on the patio barbeque affirmed that the privation of the Great Depression and war was over. The proliferation of inexpensive, drive-through fast-food hamburger restaurants also boosted beef. Beef consumption surged into the mid-1970s as the middle class grew due to government spending on education, infrastructure, and the military-industrial complex. Yet the 1970s also brought challenges for beef. Medical studies exposed a link between red-meat consumption and heart disease, which compelled many consumers to cut back on beef.

An ongoing barrage of health concerns related to the American diet gained traction after World War II, especially relating to fat and sugar. The link between red-meat consumption and heart disease was especially troubling. In 1977, the US Senate issued a report titled "Dietary Goals for the United States," warning Americans about overeating and highlighting the problem of excessive meat and fat consumption. Senator George McGovern of South Dakota, the principal author of the report, urged Americans to eat less red meat.[45]

The biggest beneficiary of the red-meat scare was poultry, which soon surpassed pork and became America's leading meat by the 1990s. The prominence of chicken was the result of technological trends that reinforced its place as a special status food. For decades, chicken was reserved for Sunday dinner, company, or courtship meals. After the war, the "Chicken of Tomorrow" program carried out by USDA researchers, grocers, and farm groups resulted in a rapid transformation of the bird. Aggressive breed improvement, the use of subtherapeutic antibiotics to

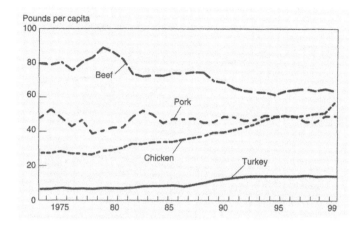

**Figure 34**

After World War II, growing affluence and concerns over dietary fat led many Americans to reject pork as an inferior meat. This graph, published in 2000, shows the challenge that pork producers faced from America's love affair with high-status beef and chicken. Source: *USDA Agriculture Fact Book, 1999*. Public domain.

promote rapid growth, and new, scientifically calibrated rations yielded a chicken that possessed a far greater portion of the carcass as breast meat with less feed and more rapidly than was previously possible. The short life cycle of chicken aided the rapid transformation. By the 1980s, families could obtain inexpensive status meat that dietary experts viewed as healthier than beef and pork.[46]

Pork and lard were special targets for fat-obsessed Americans. Lard was in decline even before World War II, but that trend accelerated after the war. Lard consumption dropped even more precipitously than pork, from annual per capita use of 13.6 pounds in 1940 to 9.4 pounds in 1959. As early as 1955 one industry insider commented that "the consumer has advised Mr. Grunt in writing—on the cash register slips of retail markets—that his hams, loin roasts, chops, shoulder butts, picnic shoulders and bacon are too fatty."[47] By 1990, the average American ate just 2.2 pounds of lard every year. The decline in lard did not mean that Americans consumed less fat; rather, they ate more. The difference was that less of it was from animal sources and a growing percentage was in the form of vegetable oil (mostly soy, canola, and corn), often in hydrogenated forms.[48] How did the pork industry respond to concerns about fat? The editors of *Country Gentleman* urged readers to do a "Major Job Remodeling Hogs," characterized by less fat on the outside and inside of the animal.[49] Carroll Plager of Hormel explained, "The meat type hog was not invented by either the producer or processor, but rather by the housewife, even though hog

**Figure 35**
This Poland China boar, photographed in the early 1950s, shows the traits that characterized much of the American swine herd before World War II. These rounded, compact animals had deep sides to maximize bacon on the carcass. By butchering time, they had a layer of fat two or more inches thick on their backs. Source: Ag Illustrated. Used with permission.

type is about the farthest thing from her mind as she does her critical shopping."[50] Over a decade later, Oscar Mayer's marketing director lamented that the pork industry had generally failed to heed what he labeled the first law of marketing: "Make what people want to buy; don't try to sell them what you happen to make."[51] If lean meat was fashionable; then producers and packers planned to deliver lean animals.

The goal was to reduce the amount of back fat on the carcass. Less back fat meant not only a greater percentage of lean meat on the carcass but also less fat in the ham and loin, the most valuable pork cuts. In other words, farmers could create *Pork People Like*, the title of a 1956 educational film developed by the University of Illinois Extension Service for hog farmers. Selling meat-type hogs would result in a better check from the packer.[52] Iowa State College communicated the same message in the film *The Pig and the Public*, emphasizing the role of women in shaping market demand for lean meat and the potential profit for hog farmers in developing meatier hogs.

State swine testing stations, first introduced in Ohio in 1954 and widely emulated, were vital institutions in remaking the hog. Farmers could board young boars at the station, which tested for feed conversion rates and back fat thickness. Station workers used a small, stainless steel ruler with a sliding gauge known as the back fat probe to measure back fat thickness on live animals. The probe, developed by Iowa State faculty member Lanoy Hazel, solved the problem of how to assess carcass quality of live animals. By making three incisions with a scalpel along the

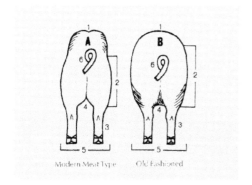

**Figure 36**

Farmers, meat packers, and agricultural experts redesigned the hog to maximize muscle and minimize fat, as seen in this comparison of a modern, lean, "meat-type" hog and a fatty, "old-fashioned" hog. Source: Roger Hunsley, "Livestock Judging and Evaluation" (AS-388, Cooperative Extension Service, Purdue University Lafayette, IN). Used with permission.

hog's back (over the shoulder and the first and last lumbar vertebrae) and inserting the probe to the depth of the connecting tissue, scientists could then measure, record, and compare back fat thickness across a population. The animals with the least back fat and best conversion ratio survived as breeding stock. Competitive hog shows awarded prizes to the best modern, lean animals, and meat packers paid premiums to farmers who delivered meaty animals.[53]

The industry succeeded in making the meat-type hog. In the early 1950s, a 180-pound hog carcass carried 35 pounds of lard. By the early 1970s, the 180-pound carcass yielded only 20 pounds of lard. Between 1963 and 1983, the average amount of fat per ham declined from 7.5 percent to 5.4 percent, while the decline in pork loin fat was from 11.4 to 7.5 percent. This reduction occurred at the same time that average carcass length increased, the size of loin eye increased, and ham and loin constituted a greater portion of the carcass.[54]

The "new pork" from meat-type animals required consumer re-education. A film titled *The New Pork: An Exciting Taste Treat* informed consumers that pork contained 57 percent less fat and 36 percent fewer calories per ounce than the old pork, claiming, "It's almost a completely

**Figure 37**
After World War II, American breeders imported the Landrace hog from Denmark. The Landrace was a longer, leaner animal than its American cousins, which helped transform American hogs from lard to meat type. This photo was taken circa 1960. Source: Ag Illustrated. Used with permission.

new meat."[55] That message of newness appeared in an ironically titled cookbook, *The American Heirloom Pork Cookbook*, that touted the virtues of new pork while emphasizing the "heirloom" nature of pork cookery.[56]

New pork was a production triumph but a marketplace problem. Cooks who grew up fearing the trichina parasite had to relearn how to cook pork. Killing the trichina worm required cooking to an internal temperature of 180 degrees Fahrenheit, but new, low-fat pork chops cooked to such high temperatures were dry and flavorless. The Iowa Porkettes, the women's arm of the Iowa Pork Producers, informed consumers that the new pork could be prepared safely at lower temperatures. Even then, new pork sometimes exhibited undesirable traits, together known as PSE for pale, soft, and exudative (watery).[57]

Some industry experts contended that consumers were asking for something that they did not really desire. Less fat, after all, meant less taste. Professor J. W. Cole of the University of Tennessee reported that while consumers voiced concerns about "greasy" pork, they also indicated that pork with marbling was preferable to that without marbling.[58] Numerous surveys, taste tests, and visual tests conducted during the 1950s suggested that consumers expressed a desire to reduce fat, but they did not always select the leaner cuts unless the differences were discernible to the naked eye. A significant minority of consumers continued to select fattier cuts.[59]

Frustrated hog producers, having successfully produced meatier hogs while failing to boost consumption, rebranded new pork. The National Pork Producers Council (NPPC) developed the "America, you're leaning on Pork" marketing theme in 1982, reinforcing the significance of pork as well as its value in a weight-reducing diet. In 1987, the NPPC's "the other white meat" campaign attempted to co-opt the presumed health benefits of chicken. They based the claim that pork was white meat on research that indicated that the substance in meat that determines color, myoglobin, was present at approximately the same levels in pork as it was in fish and chicken. The claim distanced pork from red meat and its association as a health risk. New pork was not only leaner than the old pork but also no longer a red meat.[60]

Swine industry experts cautioned in the 1990s that the lean hog was at risk of becoming too lean for consumer taste. As one representative of the NPPC stated, "I think we went way too far."[61] Maynard Hogberg, professor of animal science at Iowa State University, stated in 2005 that producers made a "leaner hog with more muscle. But they lost the taste, they lost the juiciness, they lost the moisture content."[62] Food writer Edward Behr observed in 1999 that removing the marbling from pork also removed the taste. Overcooking lean pork resulted in dry and tough meat, but marbled pork was "much more forgiving," Behr noted.[63] By the early twenty-first century, the industry compelled consumers to buy ultra-lean pork because, as journalist Nathanael Johnson observed, they could select only "what the industry chooses to give them."[64]

Consumers had expressed contradictory wishes; they wanted flavorful meat with no animal fat, but the pork industry delivered only the latter, not the former. Removing fat meant removing flavor. Although pork was leaner than ever, it still lost ground to chicken. Demand for high-fat preparations of chicken increased after McDonald's introduced the Chicken McNugget in 1983. Inexpensive nuggets, tenders, and fried chicken breast sandwiches, all cooked in vegetable oil, became the principal preparation of chicken for many Americans during the late twentieth century, undermining the presumed health benefits of chicken.[65]

### REHABILITATING PORK AND LARD

The growth of heritage pork was a consumer backlash to the ultra-lean pork. Heritage pork, characterized by a coating of back fat and a high degree of intramuscular fat, generally appealed to wealthy, urban consumers, often in markets far from the major pork-producing regions. Demand for hogs raised on pasture without gestational crates and subtherapeutic antibiotics increased after lean hogs became the norm. The appeal of heritage pork has likely broadened thanks to popular cooking shows and the increasing number of food enthusiasts.[66]

Specialty meat producers used traditional and rare breeds that mainstream farmers abandoned in the race for lean meat. The Berkshire breed, reputed to excel at converting feed into well-marbled pork, was among the most valued. In Seattle, Armandino Batali used pasture-raised Berkshire hogs from Kansas to make culatelli, an Italian-style ham sold at his store located in Seattle's Pioneer Square. Batali's culatelli was noteworthy due to the care and expertise in processing the meat and the significant amount of marbling in the ham, far exceeding that normally found in hams in the grocer's case. Herb and Kathy Eckhouse used Berkshire and Tamworth hogs to make internationally recognized prosciutto at La Quercia, located in Norwalk, Iowa.[67] Reflecting regional culture, Sean Brock of Charleston, South Carolina, reinvigorated his version of southern cooking by utilizing hogs from Ossabaw Island, Georgia. Ossabaw swine descended from hogs brought by Spanish explorers in the sixteenth century. Having survived as an isolated feral population, these animals excel at storing fat, yielding more fat on the carcass than any other breed. The fattier pig, Brock claimed, could restore proper flavor to southern food.[68] Carl Blake, an Iowa farmer, crossed a German breed of pig with a Chinese breed to yield a pig that produced fat that, according to one chef, tasted "like olive oil."[69]

Pork from the Hungarian Mangalitsa hog commanded even greater premiums than that from English and American breeds. Food writers, chefs, and gastronomes described the flavors of these specialty meats in poetic terms, invoking the language of wine tasting. Mangalitsa lard was described as luscious, while the meat "is marbled, and the fat dissolves on your tongue—it's softer and creamier." California's celebrated French Laundry restaurant featured Mangalitsa pork. However, it is rare and beyond the reach of most consumers.[70]

The most successful example of redefining pork and upselling farming practices is Niman Ranch. According to Paul Willis, ranch manager, hogs raised by Niman growers ranged from 48 to 51 percent lean, while mainstream hogs were somewhat leaner. Unlike the boutique pork sold directly to restaurants or through specialty shops, Niman successfully marketed through the Whole Foods chain, putting the product closer to the people who were most likely to demand flavor, reject aspects of industrial production, and pay more for it.[71]

Regardless of how farmers raised pork, it remained an important meat in the United States, albeit firmly in third position behind chicken and beef. Pork fell a long way from the number one position in 1944 and a high of eighty-one pounds consumed per person that year. By the 1980s, American per capita pork consumption stabilized, averaging fifty-one pounds per year. Starting in 1990, beef held steady at approximately sixty-five pounds. Chicken, by contrast, continued the postwar surge, increasing from just less than thirty pounds per capita in 1960 to eighty-two pounds by 2003.[72]

National averages, however, obscured long-standing regional, social, and cultural differences. In the twenty-first century, rural Americans continued to eat more pork than their urban and suburban counterparts did. Midwesterners led the nation in pork consumption, averaging fifty-eight pounds per capita. The preeminence of the Midwest as the leading pork consuming region should not be surprising given its prominence as the leading pork-producing region. Across the nation, men, especially those in their forties and fifties, ate far more pork than women did. African Americans consumed the most pork of any group, averaging over sixty-three pounds, while Hispanics consumed the least of any ethnic group. While the Hispanic population grew in every region, the rapid growth of this group in the West might explain the low consumption rate of pork there.

While lard was a major source of fat in Mexico, those who immigrated to the United States rapidly gave it up, substituting vegetable oils, margarine, and butter, just like Anglo-Americans.[73]

After decades of stigmatizing all fats, however, scientists began to distinguish between the threat of saturated fats and the benefits of unsaturated fats. The use of saturated trans fats in manufactured food expanded rapidly, displacing traditional sources of cooking fats such as lard, tallow, and butter. Per capita consumption of added fats and oils in the United States increased from 44.6 pounds in the 1950s to 53 in the 1970s and to over 65 in the 1990s. In 2000, the amount of added fats and oils per capita in the American diet was 74.5 pounds.[74] The substitution of saturated fat for unsaturated fat such as lard, many scientists reasoned, contributed to America's obesity problem.

Lard, it appears, is back. American nutritionists belatedly recognized that fresh lard, with a high percentage of unsaturated fat and less saturated fat than the oil that replaced it, could be part of a healthy diet. As one gastroenterologist stated, "Lard's not a big deal. The real danger in the human diet is in total calories consumed."[75] This observation is reinforced by the fact that obesity in America has worsened at the same time that the amount of animal fat in our diet decreased, leading scientists to question the place of the ubiquitous trans fats, salt, and sugar in the diet rather than simply blaming animal fat. Those looking for the causes of American obesity in the late twentieth and early twenty-first centuries could not blame lard or the high-value cuts of pork.

## PORK: FROM CIVIL RIGHTS TO SOUL FOOD

Pork remained an important food for rural southerners during and after World War II. During the war years, rural southerners ate more pork than any other group in the nation, relying on fried, fatty pork for breakfast and dinner. The war brought some changes to the South, especially as more people moved to cities to work for the defense industries. Rural southerners who moved from the countryside to Pascagoula, Mississippi, for war work consumed less fat pork and more beef than they had previously. Once in the towns and cities, they purchased beef at fast-food restaurants. The spread of national fast-food chains that featured beef in the postwar period contributed to the gradual displacement of pork.[76] The

South was not static, and the trends of urbanization and standardization altered southern foodways.

For African Americans, however, pork retained a large place on the plate for a longer duration than it did for other southerners. Pork was a prominent dish when civil rights leaders gathered at diners and restaurants to relax or strategize.[77] For civil rights activist Fannie Lou Hamer of Ruleville, Mississippi, pigs were tools to escape poverty. In 1962, Hamer underwent a transformation from farm laborer and domestic worker to civil rights activist. She understood that a secure food supply would not only improve the health of impoverished people but also strengthen the position of civil rights advocates in their communities. In 1968, Hamer organized a pig bank in Sunflower County, Mississippi, with funds from the National Council of Negro Women. She purchased five boars and thirty-five gilts and loaned bred gilts to any family that would raise a litter and, in turn, give one or two of the impregnated females to friends or neighbors. The borrower then returned the original pig to the bank. When the program ceased in 1973, over nine hundred families had participated.[78]

Some members of the emerging Black Power movement openly celebrated the food that had fueled the civil rights movement, labeling it soul food.[79] The Student Nonviolent Coordinating Committee (SNCC) linked the traditional southern diet with the Jim Crow oppression experienced by African Americans and viewed it as a tool for unity. There was a gulf between sympathetic whites and blacks because whites could not, among other things, "relate to chitterlings, hog's head cheese, pig feet, ham hocks, and cannot relate to slavery, because these things are not a part of their experience."[80] SNCC leaders glossed over the fact that soul food was traditional fare for many rural Americans, but by making their claim, they reinforced group solidarity. Soul food as a cultural marker did not escape the attention of the Nation of Islam. Established in the 1930s, the Nation of Islam gained popularity in northern cities after World War II. Unlike Black Power advocates, however, the Nation of Islam rejected pork on both religious and cultural grounds. Elijah Muhammad urged his followers to avoid pork, collard greens, biscuits, and black-eyed peas because those foods were part of a "slave diet," not the diet of free people.[81] Both Black Power advocates and the Nation of Islam agreed that soul food and pork were long-standing reminders of southern identity, although they disagreed about whether continuing historic consumption patterns was in their best interest.

While African Americans disagreed about the role of traditional foodways, the cultural hearth of soul food was changing. During and after the Great Migration, southern foodways diverged along racial lines, paralleling the white flight that occurred in response to school desegregation and bussing policies. An extensive study of North Carolina foodways conducted in the 1970s indicated the continued significance of pork in the southern diet. All families viewed pork as the proper seasoning when cooking greens, but African American families tended to season vegetables more often with ham hocks and salt pork than white families. They also fried more food in lard than white families, even those from similar income levels. The largest difference, however, was in what kinds of pork were on the table. White families increasingly consumed more high-status cuts such as pork chops, ham, and roasts. African American families ate those cuts too, but also consumed tail, neck bones, chitterlings, head, and feet.[82]

Soul food gradually became more important as a symbol than as sustenance. In 1969, a professor of nutrition concluded that African Americans who left the South continued to rely heavily on pork, lard, and other traditional southern foods such as boiled greens, sweet potatoes, corn, and legumes. The longer they were away from the South the less soul food they ate.[83] In San Diego, California, African American families seldom served pig's feet, chitterlings, and ham hocks by the 1990s, while other foods such as fried chicken and yams were on the table at least once a week.[84]

The reality was that the culinary experiences and gastronomical expectations of all Americans broadened when the United States became a middle-class nation. Food expert and cookbook author Jessica Harris observed that over the course of her residence in Brooklyn in the late twentieth and early twenty-first centuries, the dietary range of her African American neighbors expanded, paralleling the growth of the African American middle class. Describing the range of foods available at her neighborhood supermarket, Harris noted that the "traditional Southern diet of pig and corn is still consumed, but increasingly it has become celebration food for many families, eaten only on Sundays, on holidays, and at family reunions."[85]

Even the value of soul food as a cultural symbol has faded. Pastor and scholar Renee McCoy described mealtime at African American family reunions and twenty-first-century sensibilities. At her own reunion in Michigan, the slow-roasted pig was always "the star of the show." Family members brought out the cooked hog, offered the blessing, and opened the carcass. "For those of us older than thirty-five," McCoy stated, "this was a beautiful sight. The younger folks, however, thought it was absolutely disgusting." Comments and complaints about hog guts resounded in the picnic shelter. Elderly family members, "shocked and saddened by the comments," stopped the meal. One woman told the youngsters to hush and then explained the importance of hogs, rural life, and hard times. Family elders told stories "about working to raise pigs, only to be allowed to eat just the parts that had been discarded." For McCoy, the cultural disconnect was a teachable moment. Elders explained how "the pig represented an indispensable partner in the African American quest for life," affirming the value of the family reunion, intergenerational sharing, and southern roots, not to mention a traditional meal, innards and all.[86]

## THE UNITED STATES OF BACON?

If soul food has faded in importance, fat pork products have not, despite the campaign to disassociate pork from its fatty history by remaking it into white meat. The consumption of processed pork products such as bacon and sausage outpaced that of the leaner fresh pork cuts. Americans who ate pork at home consumed slightly less than 20 percent of their pork as fresh cuts such as chops and roasts. The rest was as ham (31 percent), sausage (20 percent), bacon (18 percent), and processed lunch meats (10 percent). Ham was the most popular processed pork consumed at home, with sausage and bacon in second and third positions, respectively.[87]

Tinned pork products such as Spam were exempt from the trend toward more processed meat, with consumption stagnant. One Vietnam War veteran recalled that his mother sent Spam to him in care packages. It was, he fondly recalled, "like gold."[88] Nostalgic Americans purchased kitschy Spam T-shirts, coffee mugs, toys, and other products from Hormel's online store. In Waikiki, Hawaii, a street festival called the Spam Jam celebrated the significance of Spam in the local diet. Hawaiians consumed an average of five cans per person, per year, more Spam on a per capita basis than any other group of Americans. From 1995 to 2007, a group from Austin, Texas, organized the Spamarama festival, which featured

**Figure 38**

As pigs became leaner, Americans increasingly craved fattier cuts and preparations. This is a photo of pork meatloaf wrapped in bacon, a recipe promoted by the National Pork Board. Source: National Pork Board. Used with permission.

cooking competitions and Spam-themed events. Despite the nostalgic turn, Spam was held in such low esteem that it became the label for unwanted or pernicious email.[89]

Bacon, by contrast, was America's most high-profile processed pork product. From the 1980s through the 1990s, bacon was cheap. As one industry leader claimed, pork belly was a financial "drag on the carcass" due to concerns about dietary fat and cholesterol. However, in 1992, Hardee's introduced the Frisco Burger that featured bacon, inaugurating a fast-food bacon race. Chains topped burgers and fried chicken sandwiches with bacon to revive interest in the aging franchises and to maintain position in a crowded market.[90]

The restaurant marketing campaigns of the 1990s that promoted two strips of bacon on a burger look tame today. In early 2014, Carl's Jr. and Hardee's introduced four strips of bacon on burgers and breakfast sandwiches.[91] Bacon has even accompanied ice cream, with a bacon sundae available at Denny's restaurants in 2011. Burger King developed its own bacon sundae in 2012, and Jack in the Box restaurants offered a bacon milkshake in 2012.[92] In early 2015, Little Caesars restaurants introduced a deep-dish pizza with 3.5 linear feet of bacon wrapped around the crust in addition to toppings of bacon and pepperoni. A market researcher observed that bacon had become "the ultimate indulgence" for consumers, evoking emotional connections as a comfort food.[93]

Given the expansive place of bacon on American menus, it is not surprising that consumers celebrated it at festivals from coast to coast. In Iowa, a group of friends gathered at Spirit Lake in 2001 for a weekend getaway to celebrate bacon. In 2008, their private celebration became public when they organized the Blue Ribbon Bacon Festival in Des Moines. The event featured local restaurants and vendors with sample food items, bacon-themed entertainment, competitions, and merchandise. Beginning with just a few hundred participants in the first year, the Blue Ribbon Festival sold out ninety-five hundred tickets to the event in 2013 and hoped to sell twelve thousand in 2014. Leo Landis, aka the Bacon Professor, is a featured speaker. Landis travels the breadth of the state to sample the products of rural and small-town lockers, discussing and presenting the finer points of making and tasting bacon. A group in Chicago organized Baconfest in 2008, with other Baconfest events held in Washington, DC, and San Francisco. Zingerman's Delicatessen of Ann Arbor, Michigan, has hosted Camp Bacon since 2010. Dozens of communities from Richmond, Virginia, to Portland, Oregon, jumped on the bacon festival bandwagon.[94]

Bacon was also omnipresent on television. Cooking shows featured bacon prepared in a variety of foods, cooked every conceivable way. The *United States of Bacon* debuted on the Discovery Channel in 2012. The program host visited restaurants around the country that featured distinctive preparations of bacon. A writer who described the new show suggested that having moved to the small screen, the next frontier for bacon was a feature-length movie.[95] In 2014, that actually occurred. The mockumentary film *State of Bacon* featured the Blue Ribbon Bacon Fest and was filmed at the Des Moines event and other central Iowa locations. While elite chefs reported a degree of bacon fatigue among the staff at high-end restaurants, most Americans showed no signs of slowing consumption.[96] The United States, in other words, has not reached peak bacon.

Bacon and other processed pork such hot dogs, lunch meat, smoked ham, and sausage represented the largest growth area in recent American consumption. The popularity of processed meats mirrors the popularity of "healthy" chicken that consumers enjoy mostly as fried tenders and nuggets. Health-obsessed Americans, after all, still like salty and fatty preparations. The growing importance of processed and preserved meats represents a bit of déjà vu. The old preserved meats that were so popular for laboring people, free and enslaved, became fashionable in a predominantly middle-class nation and among elite gastronomes. With bacon's resurgence, prices have increased accordingly. Bacon is most often affordable on fast-food burgers, continuing an association with low-income Americans, but it is increasingly a luxury often beyond the reach of working people.

CHAPTER 8

# SCIENCE AND THE SWINEHERD

In 1962, Iowa State University opened its "Pigneyland" exhibit at the Farm Progress Show, an annual three-day event held that year on a farm in Hamilton County, Iowa. The exhibit evoked Walt Disney's already iconic Disneyland to introduce midwestern farmers to the latest in ideal market animals and production techniques. Pigneyland as "Tomorrowland" was a place of concrete, metal slotted floors, metal pens, and farrowing stalls as well as flesh. It featured four buildings, or lands, each dedicated to teaching farmers about a different aspect of production: carcass quality, breeding selection, farrowing-starting, and growing-finishing. Experts advocated for distinct facilities for different phases of production and carefully calibrated rations for each phase of the life cycle. A refrigerated trailer at the exhibit displayed ideal carcasses and pork cuts to show what "hog men should aim for" when attempting to meet market demand.[1]

Instead of a revolutionary break with tradition, however, the Pigneyland exhibit and the growth of confinements and concentrated animal feeding operations (CAFOs) represented evolution in the American livestock experience and American business. During the second half of the nineteenth century, agricultural reformers applied scientific techniques to feeding, breeding, housing, and health in the name of efficiency and profit. Animal husbandry experts had long emphasized the efficient conversion of feedstuffs into flesh, careful management of housing and health, labor saving, and rationalization of animal production to take advantage of favorable market conditions. Farm experts conceptualized the hog as a machine as early as the nineteenth century, and it was a modest step to reorganize production around new technologies consistent with the ideal of industrial efficiency. Even though the size and technology of large-scale, twenty-first-century operations would have appeared staggering to those who visited Pigneyland in 1962, they spoke the same language of production.

## MANUFACTURING HOGS: FROM OPEN RANGE TO PASTURE AND PEN

Across much of the South at the end of the Civil War, free-range hog husbandry was the norm, with scant attention to breed improvement and husbandry. An 1875 survey of Georgia farmers revealed the state of hog husbandry in Georgia based on several hundred replies to a circular. Thomas Janes compiled the responses, reporting that many Georgia farmers simply marked their pigs and turned them out in the woods "feeding, perhaps, once a day, just enough corn to keep them from growing wild." Almost two-thirds of all Georgia correspondents reported that "neglect, want of food, proper management, and good fences" constituted "the principal difficulties in the way of success." Not surprisingly, with such a hands-off approach, over a third of the respondents noted that thievery was the most important obstacle to hog raising in their respective neighborhoods. The Georgia survey was incomplete, but the replies suggested that open-range husbandry remained popular and that there was little interest in change among many yeoman farmers.[2]

Some southern farmers continued to raise hogs in the woods "year round without feed or care" as late as 1900. One northern traveler described the use of a "hog dog" that was trained to seek out a herd of hogs in the woods, bay at them, and then allow the pigs to chase it in the direction of home. The dog led the animals into a log pen with a fence low enough for the dog to jump but high enough to contain the hogs. The lean woods hog or razorback, according to this writer, was "enterprising . . . intelligent, suspicious . . . a born fighter, a good rustler, shifty as a New England Yankee, and courageous as a confederate colonel." Such animals, he asserted, would "hold the country until fenced pastures, tame grasses and cheap corn invade it."[3]

Despite the prevalence of the open range, the practice was under siege. Increasing population density and the desire of political leaders to exercise more control over land and people hastened the end. As one historian noted, attacks on subsistence farming and the commons in the name of modernization helped the Democratic Party reclaim control over state politics from "'traitorous' white Republicans and their 'colored' allies."[4] The problem of hog theft indicated the intensity of political, class, and racial tensions of the period. Historian Steven Hahn explained that white elites believed closing the range would increase the dependency of poor African Americans and keep them tied to the plantation and commodity production. Closing the range was palatable to poor whites who relied on subsistence production because state laws allowed for local control rather than statewide mandates.[5] Promoters of local option laws at least paid lip service to the rights of white yeomen farmers.

The combination of the closing of the range and the rise of cotton sharecropping was bad news for many southerners. Self-sufficiency faded as poor farmers who lost the range needed to keep low-cost livestock. In a six-county section of Mississippi and Alabama's major cotton-producing region, the numbers of hogs per capita fell well into the twentieth century.[6] As a result, swine became less important for farmers on the economic margins after the Civil War.

For most northern farmers a mix of pasture and penning prevailed. Iowans passed a series of fence laws with a local option in the early 1870s. Northern opposition to fence laws was on the same grounds as in the South: that the small or struggling farmer needed cheap livestock feed. Those feeders who simply turned their animals loose and took their risks were, by 1900, "crowded out," as an Indiana farmer observed. He stated that make-do farmers were replaced by those "who prepare not only an abundance of feed, but also the best of shelter for their flocks and herds throughout the entire winter."[7] In 1899, a Kansas farmer complained that

**Figure 39**

In 1870, Joseph Harris published his version of the ideal form of market swine. "The nearer he will fill a rectangular frame," Harris asserted, "the nearer he approaches to perfection of form." Of course, not all hogs met this standard during the late nineteenth century, but for Harris and other promoters, this shape allowed for maximum space for the vital organs that, in turn, produced the best quality and quantity of meat and lard. Source: Joseph Harris, *Harris on the Pig: Practical Hints for the Pig Farmer* (New York: Orange Judd, 1870). Public domain.

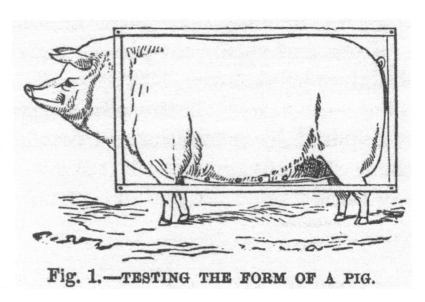

Fig. 1.—TESTING THE FORM OF A PIG.

there were still farmers who believed that "a hog is a hog no matter if it is the worst kind of a scrub." He noted that razorback-style hogs were not profitable and "one had better be without them."[8]

Improvements in breeding and husbandry gained traction after the Civil War, although change occurred more rapidly in the northern states that emerged mostly unscathed by the Civil War. Jacob Biggle, author of *Biggle Swine Book: Much Old and More New Hog Knowledge, Arranged in Alternate Streaks of Fat and Lean* (1899), recognized that "hog husbandry is undergoing changes." He asserted, "Experience has heretofore been the main guide but science now comes to the swineherd's aid."[9] Biggle relied on traditional folk practices but also used new scientific knowledge to make pig farming more profitable.

Northern farmers were in a better position to embrace improved breeds and reject what progressive farmers derisively labeled the razorback, land pike, or woods hog. The new ideal was less long and lean and more square and blocky. In 1870, Joseph Harris depicted a hog body in a rectangular frame, with only the head and legs protruding. Harris spoke for all improvers when he asserted that the nearer the animal filled the frame, "the nearer he approaches perfection in form."[10] Subsequent manuals repeated the same guidelines. In 1900, a USDA manual proclaimed that the "outline of a perfect hog, when viewed from any direction, should be nearly that of a parallelogram with the corners slightly rounded."[11] This rule applied to all the breeds, suggesting that Berkshire, Poland China, or Duroc-Jersey could all conform to a common ideal. As the director of the Mississippi Agricultural Experiment Station cajoled, farmers who would not purchase a pureblood boar and sow should at least purchase a new pureblood boar to breed up the herd.[12]

**Figure 40**

Castrating young boars was critical to ensure controlled breeding once farmers abandoned the open range. Farmers who invested in purebred boars wanted to prevent grade animals from impregnating sows. Farmers also found that castrating males prior to sexual maturity improved the flavor of the meat and prevented aggressive behavior. Source: Ag Illustrated. Used with permission.

The number of purebred hog breeders proliferated, with each one anticipating a windfall from the sale of blooded stock in the name of more efficient fattening. Midwesterner Silas Shepard, the author of an 1886 treatise, contrasted the traditional hog raiser with the new hog breeder. The raiser was "content to plod along the dull, dusty road of the thoughtless past and leave Nature as he finds her, to do as his father did." To Shepard, the breeder was a figure to admire. He was "in a certain sense a creator" who transformed scrubs into new breeds. "Life is too short," Shepard claimed, "to waste in feeding scrubs."[13]

If breed improvement was the foundation for profitable pig farming, what kind of advice did experts provide for improvement? The first job was to do more crossbreeding. This was contrary to some breeders' views that only purebred animals were worth feeding. Joseph Harris contended that maintaining pure breeds was necessary for the improvement of the species, but purebred animals did not make the best market animals. Here he drew on the experiences of English farmers as well as his own practical experience as a purebred breeder. Crosses of purebred animals with common or native animals combined characteristics that buyers prized.[14]

Farmers who wished to produce purebred animals faced the problem of maximizing investment in a blooded boar. Some elite breeders no longer tolerated simply turning the boar out with the sows and hoping for copulation and conception. One technological solution was the breeding crate. The crate was open on one end, allowing the sow to enter, and then closed in from behind. The boar could then mount the sow with supports on either side of the box to help bear the boar's weight. Breeding crates, developed in South English, Iowa, as early as 1896, gained popularity in the 1910s. Historian Gabriel Rosenberg claimed that crates were in widespread use among breeders of blooded stock by 1930.[15] Even so, only a small minority of hog farmers committed to the purebred business.

In the late 1800s, references to swine as machines to convert grain into flesh were common. Joseph Harris observed in the 1880s that "a pig is a mill for converting corn into pork."[16] In 1900, one farmer titled an essay "Manufacturing Hogs," connecting the processes of farm and factory. Manufacturing required good management to make a profit, and this farmer urged his peers to "push from start to finish and put [hogs] on the market at once."[17] William Dietrich, a University of Illinois swine expert and author of *Swine: Breeding, Feeding, and Management*, compared the process of feeding and maintaining a pig to the operation of a steam engine and threshing machine. Hogs on the Peterson farm in Hardin County, Iowa, constituted "a factory" that transformed low-cost grain, forage, and supplements into valuable products, according to a writer for *Wallaces' Farmer and Iowa Homestead* in 1917.[18] Such imagery employed by farm experts and journalists signaled to farmers to conceptualize living organisms as economic units, even machines.

Management of diet, housing, and sanitation was central to making pork in an increasingly mechanized regime. For most hog farmers, pasture continued to play a critical role in hog management through World

**Figure 41**

Pasture played a critical role in hog farming through the mid-twentieth century. Experts advocated grazing on protein- and calcium-rich alfalfa pasture during the summer months for best weight gain and optimal health. This photograph is from 1916–1918. Source: Ag Illustrated. Used with permission.

War II. From spring to fall, hogs in the major swine-producing areas spent much of their lives on pasture. One writer stated that large-scale hog farmers could not make a profit on hogs without pasture.[19] Summer pasture was "the secret of success" according to Foster Coburn, the author of *Swine Husbandry*, first published in 1877 and reprinted in 1897. Animals on pasture developed strong skeleton and muscle, preparing the animal for the round of fall fattening. Good pasture was an article of faith and science during this period. As one writer noted in 1870, "The farmer who proposes to make money by raising pork, must have a pasture for his swine during the season of grass."[20] The editor of *Prairie Farmer* observed that in Oregon, most breeders regarded clover to be "indispensable."[21] Coburn contended, "The necessity of providing swine with summer pasture and green food, is, even in the best corn-producing districts, become more and more apparent."[22] Farmers put hogs on pastures of clover, timothy, bluegrass, and rape (later known as canola). In midsummer, wise farmers used supplemental feed to offset dry conditions and poor pasture. Reformers cited English experiments with peas and urged greater use of them in the United States. Turning hogs into recently harvested fields of small grain such as oats, rye, or wheat to glean in midsummer was an alternative to overgrazing pastures.[23] Southern farm reformers prescribed livestock and forage crops as a solution to southern farm problems. In a USDA bulletin from 1899, the former director of the Mississippi Agricultural Experiment Station encouraged southern farmers to utilize pastures for growing hogs and to rely on a diversity of feeds, including sorghum, peanuts, and artichokes, not just corn.[24]

Corn remained the most valuable fattening ration, distributed through self-feeding, dry-lot feeding, or hogging down. Iowa State's John Evvard claimed that "any class of swine on the farm which are ready for fattening can be successfully turned into the cornfield."[25] Hogging down was especially valuable during periods of labor scarcity, such as wartime. As one writer counseled in 1918, September was the time to "put the pigs to work and enjoy the fruits of the swine labor."[26] Hogs fed in drylots gained more efficiently than those in the cornfield did, but many farmers justified hogging down because of labor savings, especially during the wartime emergency.[27]

The subject of fattening hogs continued to attract the most attention from farmers and researchers at the state agricultural experiment stations. Joseph Harris reported on feeding experiments conducted at Michigan Agricultural College as well as those conducted in England in 1886. The English study focused on determining the proper kind of fattening ration. Harris noted the author's suggestion that breeding was more important than feed quality in achieving optimal weight gain, reinforcing the role of blooded stock in improving herds.[28]

English consumers, however, wanted leaner hogs with more meat in the bacon, which meant less corn. Foster Coburn asserted that feed quality mattered to yield the "most healthful, most palatable, and most eagerly sought" pork with the right mix of fat and lean and in the most economical manner. Experiments at the Missouri Agricultural College and the University of Wisconsin indicated that a mixture of corn with shorts (coarsely ground grain with bran and the germ) and skimmed milk made a meatier and healthier hog.[29]

Cooked feeds, so popular in the mid- to late nineteenth century, were out of favor in the twentieth century. The authors of an Oregon State Agricultural Experiment Station bulletin on fattening swine cited studies conducted at Corvallis and elsewhere to argue that "cooking of grains is not only of no value but has a decidedly detrimental effect upon the feed."[30] Cooking feed was now appropriate only for sick animals or to prepare otherwise unwholesome food. One author concluded that it was "not advisable to make radical changes in the temperatures or nature of hog foods."[31]

Experts agreed that corn alone was not a sufficient ration, especially given the new emphasis on producing bacon with a streak of lean to complement the fat. Iowa State College experts argued that the best feed for hogs on drylot was a mix of corn, middlings (by-products of wheat milling), tankage (rendered animals), and salt. Soybeans were also a protein-rich feed. Progressive hog producers in the South supplemented corn with a variety of other feeds such as forage crops, sweet potatoes, and cottonseed meal.[32]

Dairy waste, where available, was popular and inexpensive hog feed. The author of *500 Questions Answered about Swine* (1907) stated that while it was possible to raise hogs without dairy products, "better hogs and cheaper hogs can be raised with it."[33] Farms located within the "milk shed" of a city (the area where it was profitable to sell milk to urban consumers) were unlikely to have skimmed milk for their pigs, but farmers beyond the milk shed who sold cream rather than fluid milk had leftover skimmed milk as a by-product.

Feed supplements were increasingly important in the first half of the twentieth century. Hog husbandmen at the Ohio Experiment Station reported that supplemental phosphorous, salt, and protein allowed hogs to gain weight with less feed. Tankage was one of the most common supplements because it was high in both protein and phosphorous. Although the merits of tankage and other supplements were widely discussed, many farmers did not utilize them. Those who did, however, did not necessarily feed a balanced ration. One observer claimed that too many farmers bought the occasional sack and fed it "like it was some kind of a tonic."[34] Some farmers reasoned that expenditures for supplements were unneeded when grain produced on the farm was already at hand. According to one farm journalist in 1942, farmers "subconsciously" regarded corn as free feed. A "good hog man," however, understood that minerals were critical to maximum gains.[35]

Farmers placed renewed emphasis on swine housing. The proper well-ventilated hog house had ample natural light and was located on well-drained land to carry waste away from the structure. Southern farmers tended to invest little in hog housing due to their comparatively mild winters, but northern hog houses were often more elaborate. Like those of the pre–Civil War period, houses included feeding and bedding areas, small runs or yards, and sometimes space for grain storage. Sleeping apartments for market hogs measured six by eight, eight by nine, or seven by twelve feet, with similar sized feeding pens. The runs were often double the size of the sleeping apartment. These houses, however, were generally for the fall fattening period and farrowing. They were not year-round homes since market hogs were to be on pasture during the late spring through early autumn.[36]

Housing prescriptions of the early twentieth century differed little from those of the late nineteenth, although many farmers neglected constructing special buildings for hogs. Complaints about housing were common throughout the period. One observer noted that on most farms "the hog house is the poorest building on the farm," often "just an old shack of a place, without windows, ventilation, or even a tight roof and walls."[37] But the increasing importance of management in farm operations encouraged better housing. As one author of a treatise of farm buildings noted in 1905, "What would be the business future of the swine breeder whose hogs were compelled to be exposed to the cold blasts of winter or the glaring sun of summer?"[38] In 1942, a speaker at the Minnesota Swine Feeders meeting explained how he could determine whether a farmer provided adequate housing for hogs without even inspecting a hog house. If he drove into a farmyard at eight o'clock on a winter's night and heard agitated hogs squealing and squabbling, he knew that they were uncomfortable and poorly housed. Comfortable hogs, he claimed, would be asleep.[39]

Ideal hog houses were on land with good drainage and had appropriately sized pens for feeding and sleeping. One central-alley Illinois structure had twelve pens measuring six by nine feet arrayed six on either side of a four-foot-wide alley. Government experts recommended that farrowing pens range from six by eight to eight by ten feet.[40] They also emphasized the significance of ventilation and sunlight, just as nineteenth-century sanitary reformers had, but the new twentieth-century houses maximized sunlight, "the Natural Disinfectant."[41] The proper location of windows depended on the height of the sun during farrowing season, maximizing the spread of sunlight indoors to promote healthy sows and pigs.[42]

**Figure 42**
These Duroc Jersey hogs on pasture consume soybeans as a feed supplement in this 1927 photograph. A barrel on top of a cast iron wheel from a hay mower mounted on a sled made an inexpensive self-feeder. Source: Ag Illustrated. Used with permission.

Moveable hog houses were especially popular. Farmers could construct small, portable sheds on runners, either A-framed, also known as the Lovejoy type, or the box type (aka Bonham), both with hinged roofs for access and cleaning. The box-type shed was eight feet square, while the A-frame was eight by six. Both could be constructed with open ends for warm climates or enclosed for areas with cold winters.[43] Portable houses located on pastures put animals directly where they would eat and lessened the likelihood that they would stand in their own waste and consume their own parasite-laden manure. As a Bureau of Animal Industry (BAI) husbandman asserted, portable pens "make it more difficult to carry contagion to all animals in the herd."[44]

Cleanliness and sanitation were critical parts of the new hog husbandry. As long as farmers left hogs on their own recognizance, there was no concern about cleanliness. As one writer noted in an 1868 issue of the *Prairie Farmer*, in the "old days, the hog was a cleanly animal," eating fresh grass, drinking fresh water, and bedding down in the woods. The writer explained, however, that "now all this has changed; the hog is kept shut up, and his former liberty is taken away" and the hog was unclean at the hands of man, not by its own doing.[45] Many farmers gave little thought

**Figure 43**

Many farmers relied on makeshift shelters for their animals well into the twentieth century, even as experts and improvement-minded farmers turned to specialized, sanitary structures for hogs. This rustic shelter was photographed in 1922. Source: Ag Illustrated. Used with permission.

to changing their own behavior even as they expected pigs to adapt to enclosures. According to Jacob Biggle, "It is time to regard the pig as a clean rather than as an unclean animal; and I think the markets will compel this change of treatment, for cleanliness is directly in the line of choice pork products."[46] Cleanliness took on new importance as farmers in the corn-growing regions kept larger herds and the risks of hog cholera emerged. In addition to cleanliness, Biggle stated, "experience teaches that the disease [hog cholera] more commonly appears in large herds than in small ones. The moral of this, then, is easily understood. Do not keep hogs in large droves. I do not believe that over twenty-five or thirty hogs at most should long remain together."[47]

Prior to 1950, every author of a hog-raising manual, tract, USDA bulletin, or article about hogs urged farmers to keep hogs on pasture during warm weather. The most typical pasture system involved farrowing on pasture either in portable houses or in a central hog house, then moving the young pigs to pasture upon weaning. Experts advised that alfalfa and clover pastures were the most desired for hogs, although Lovejoy urged farmers to move to a pasture of rape around midsummer.[48] Ulbe Eringa of South Dakota described his eighteen acres of alfalfa as "beautiful hayland and foraging for the hogs" in 1908.[49] For southern farmers, pastures of alfalfa, rye, rape, cowpea, sorghum, peanuts, artichokes, and sweet potatoes were valuable summer feed prior to fall penning.[50]

Farmers who grazed hogs invariably ringed their snouts to protect their pastures from rooting hogs. Harry S. Truman of rural Grandview, Missouri, wrote to Bess, "When a nice bluegrass pasture is at stake I'd carve the whole hog tribe to small bits rather than see it ruined."[51]

One of the most important developments of the period was self-feeding, a practice often used with hogs on pasture. Self-feeders were small wooden structures or barrels with a small trough at the bottom. The farmer filled the bin with grain and a supplement such as tankage, and as the hogs ate gravity pulled feed from the bin into the trough. It was easy to mount self-feeders on skids for convenient moving between lots or pastures. A study of self-feeding versus hand feeding indicated that the self-fed hogs used less grain to gain a hundred pounds. Furthermore, those gains required less labor. Citing these rapid gains and labor costs, International Harvester's Extension director endorsed the practice as "O.K."[52]

Cement feeding floors gained in popularity for sanitation- and waste-conscious farmers. The idea was that the hard surface would keep the animals and feed out of the muck and mire, preventing wasted grain and supplements. Furthermore, feed contaminated with manure spread infection and parasites. Some farmers began using cement in feeding floors in the late 1800s, but the practice was limited due to the high cost of materials.[53] The rise of the ready-mix portland cement industry made feeding floors more affordable, although they were still rare before World War II. In 1918, an Indiana farmer reported his transition from feeding on the ground to a feeding floor made of lumber and finally to concrete.

**Figure 44**
Farmers utilized improved hog houses for farrowing or shelter during the winter months. In this hog house, partitions were set up in preparation for new occupants. Source: Ag Illustrated. Used with permission.

He favored concrete because the wood decayed so rapidly and attracted rats. Cleaner feedlots, he reasoned, meant less disease risk.[54]

## HEALTHY HOGS: DOCTOR SANITATION, THE MCLEAN SYSTEM, AND THE SHAY METHOD

The health of the hog, asserted one Indiana veterinarian, was the key to profitability. "No animal on the farm is more susceptible to unfavorable health conditions," Dr. A. C. Spivey argued, "nor is there one that will respond more readily to favorable health conditions." While health was a product of breeding, housing, and feeding, only "Doctor Sanitation" assisted by farmers functioning as nurses could save the herd.[55] Spivey's emphasis on traditional sanitary solutions such as manure removal, sunlight, and cleanliness was readily adaptable to an approach based on germ theory.

The most notable attempt to control contagion of the early twentieth century was the McLean system. The McLean system of sanitation would improve the health of all pigs, but it was specifically planned to solve the problem of high mortality among young pigs prior to weaning. A study of farm production costs from the early 1920s indicated that farmers lost approximately 41 percent of their pig crop prior to marketing to diseases

**Figure 45**

A mixed-breed sow farrows in an A-frame structure on a pasture in Tazewell County, Illinois, in 1952. Some experts advocated farrowing on fresh pasture to minimize disease risk associated with contaminated hog houses. Source: Audio-Video Barn, Illinois State Museum. Used with permission.

such as necrobacillosis and scours.[56] BAI officials, firmly devoted to germ theory, developed the McLean system in Illinois to control worms and necrotic infection that constituted a significant drain on hog profits.

Pioneered in McLean County, Illinois, in 1919 and tested through the mid-1920s, the McLean system involved four "simple but necessary" procedures. Farrowing pens were cleaned and scrubbed with a lye solution, followed by spraying with a cresol and water solution. According to E. T. Robbins, associate professor of Animal Husbandry Extension at the University of Illinois, "hot water kills worm eggs; the lye loosens the dirt; the disinfectant destroys germs of infectious diseases." The second step was to clean the sides and udder of the sow with soap and water prior to farrowing to prevent the newborn pigs from ingesting parasite eggs. After farrowing, farmers hauled the sow and pigs to their new pasture rather than herding them to prevent picking up any parasites in transit. Finally, the young pigs remained on a clean pasture until the age of four months.[57]

The plan relied on farmer cooperation, which required a significant publicity campaign. Farmers encountered advice in the farm press, extension or USDA publications, land-grant school short courses, 4-H clubs, and public demonstrations. The railroads collaborated with state extension programs to spread the message of the McLean system and better management to help farmers reduce the number of sows needed to market the optimal number of swine on a given farm. The Burlington Railroad organized the six-car Pig Crop Special to tour Iowa and Nebraska in September 1929. The Pig Crop Special included a Pullman car and café car for staff members, one car for livestock, one flatcar with a hog-pen exhibit, and two exhibit cars. A highlight of the exhibit was Susie the Talking Sow, a mechanical sow with a speaker mounted inside that would answer questions on pig care, especially how the McLean system would help keep young pigs alive. Local organizing committees and county agents reported favorable responses from most of the public, although the barbeques, free ice cream cones, and free cigars probably helped boost interest.[58]

Agricultural reformers taught young people new techniques to change farming in the present as well as the future. If young people accepted reform messages and showed that new techniques were profitable, parents might change too. Members of youth Pig Clubs raised their own animal, keeping track of feed, pasture use, and weight gain. Criteria for prizes included best overall animal suited to the purpose (bacon or lard), daily gain, cheapest gains, and record keeping.[59]

William W. Shay, a Michigan farmer transplanted to North Carolina, was one of the most famous advocates of hog improvement. In 1918, after a decade of successful hog raising and mixed farming, North Carolina State College at Raleigh hired Shay to wean farmers from cotton dependency. As Shay stated, "the problem was to find some other crop that was equally as profitable as cotton" while reducing "the cost of producing cotton by increasing the yield per acre, which, of course, meant soil improvement."[60]

By the mid-1920s, Shay had articulated his Shay method to rectify unprofitable hog husbandry practices and boost agricultural diversity. Shay emphasized proper housing and sanitation, the importance of corn in the diet for market hogs and brood sows, and establishment of a breeding cycle that permitted hogs to reach market weight at the time for optimal prices.[61] Shay made only modest progress in changing the North Carolina hog industry, but he did show that it was possible to lessen the dependence on cotton and make hog raising profitable. Reformers like Shay drew on midwestern practices to remake southern agriculture.

The Earl Butler farm in Iowa County, Iowa, was an exemplary model of best practices that experts such as Shay would have approved. In

September 1926, the county extension agent gave a tour of leading farms for over a hundred farmers that included the Butler place. Butler used purebred Duroc sows and a Tamworth boar for crossbreeding, employed the McLean County system, pastured on fresh ground, and then fed corn and tankage after weaning. The tourists observed a "phenomenal" lot of spring pigs, weighing close to two hundred pounds each. Never before, the correspondent Henry A. Wallace noted, had he seen "such a lot of uniform, heavy, smooth spring pigs early in September."[62]

Model farms such as Butler's, however, were the exception. Twenty years after the introduction of the McLean system and the work of William Shay, many farmers practiced low-input hog husbandry and still suffered significant losses among young pigs. In the early 1940s, Iowa State Extension staff estimated that only 57 percent of all pigs survived into adulthood to reach market weight.[63] Experts urged farmers to redouble efforts to "save the pigs." Reducing the mortality rate of shoats and feeder pigs remained one of the biggest challenges facing farmers as they looked into the postwar period.

Even in the major corn- and hog-producing regions, conditions on farms varied widely. On one end of the spectrum, progressive farmers like Butler utilized purebred boars and sows to obtain crossbred animals, mostly of the lard type. They constructed hog houses and portable hog shelters, relied on a mix of pasture feeding of forage crops, drylot, and hogging down. They supplemented the ration with dairy waste, tankage, and minerals. They vaccinated against hog cholera and practiced the McLean system of sanitation to reduce disease losses. On the other end of the spectrum, farmers simply fed corn and ignored roughage and supplements. They let hogs shift for themselves or provided rudimentary shelter in muddy lots and paid little attention to the kind of hog they raised.

Most farmers occupied points along the swine husbandry spectrum rather than the extremes. A study of central Illinois farms in the 1920s revealed that many families earned good returns on hogs at the same time that others suffered losses. In 1925, the returns per hundred pounds of pork ranged from a loss of $6.78 to a profit of $6.04. The next year the range was from a $4.64 loss to a $9.01 profit. Hogs marketed in the fall glut received lower returns than those marketed in the summer, which explains some of the variation. The principal lesson, though, was that husbandry practices influenced the bottom line.[64] It is impossible to know how many American farmers made money on hogs, but the study revealed that the experiences of hog farmers varied widely even in the same neighborhood. Promoters of aligning hog production to machine standards had a long way to go to achieve their goal.

**Figure 46**
Maintaining sanitary conditions was important for keeping young pigs and sows alive and healthy. Here, members of the Klafke family of McLean County, Illinois, practice the techniques of the McLean system of swine sanitation to reduce the risk of parasites on their farm in August 1942 prior to farrowing fall pigs. Source: McLean County Museum of History. Used with permission.

## A COMING WAY TO RAISE HOGS: THE ROAD TO CONFINEMENT

Shrinking returns, changes in consumer demand, and seasonal variability in market prices were long-standing, difficult problems. Could evening out the marketing cycle provide relief? Were there efficiencies in production that could lower the cost of making a pound of pork? Could farmers save the estimated 30 percent of all young pigs that died prior to weaning and another 10 percent that died as shoats? The answers to these questions, according to the *Farm Journal* in 1956, were yes, but only if hog farmers learned to "raise pigs like broilers." Farmers who used scientific feeding and improved genetics and moved hogs from pasture onto concrete could cut labor costs, lessen the effects of disease, farrow hogs year-round instead of twice per year, and ultimately boost profits.[65] Several simultaneous developments paved the way for greater efficiency in production and marketing that enabled the intensification and increased scale of hog farming.

In the search for just the right vitamin or nutrient to maximize livestock production in the 1940s, scientists Thomas Hughes Jukes and E. L. R. Stokstad discovered that animals fed with vitamin B12 gained weight on less feed than those without B12. The real surprise was that it was not the vitamin but the source material used to produce it—residue from the manufacture of antibiotics. Stokstad and Jukes labeled the phenomenon the "antibiotic growth effect."[66] Antibiotics also aided early weaning. Sanitation had not solved the problem of high mortality among young pigs. In 1954, Iowa State College researchers developed a dry feed they called "pre-starter" to give to week-old pigs that would shorten the weaning period from eight weeks to two. The result was improved survival rates among newborn pigs. The antibiotic suppressed disease in the pigs and hastened recovery for the sow postpartum. The results were miraculous. "For every farmer who questioned the economy of early weaning," one farm journalist observed, "I found five who say it's profitable despite the high cost of pre-starter."[67]

Dr. Damon V. Catron, an animal nutrition expert at Iowa State College, facilitated the effort to raise pigs like broilers. Catron applied the lessons of the poultry industry to the hog business, emphasizing nutrients and vitamins, carbohydrates, and enzymes for maximizing production. Among his contributions to nutrition was identifying the key times in the life cycle when high-cost, nutrient-laden rations were most important and when it was possible to provide low-cost, carbohydrate-rich rations.[68]

Catron argued that greater control over the physical environments of hog production would maximize investments in quality feed and breeding stock. The key was mass production, characterized by multiple farrowing per year. Year-round production and marketing of hogs would level the highs and lows of the market and make hog farming more profitable. If farmers could move sows and pigs onto concrete and into climate-controlled buildings with natural or artificial ventilation, they could improve mortality rates, better use automated materials handling for feeding and manure removal, and improve the conversion rate of feed into flesh. In a special 1955 issue of *Successful Farming*, Catron contended that improved control over housing, diet, and breeding was "A Coming Way to Raise Hogs."[69] Ideally, hogs in confinement had less exposure to the

**Figure 47**
Detail of a portable shade structure for pastured hogs on the Stagge farm, 1964. Source: Ag Illustrated. Used with permission.

environmental stresses of heat, cold, humidity, and poor sanitation. Comfortable pigs were profitable pigs.

Despite the promised benefits, most farmers did not intend to experiment with expensive new facilities. Many were willing to confine hogs on concrete lots for feeding (the first definition of confinement), use scientifically prepared rations, and even focus on producing a meat-type hog, but total confinement was too expensive for most farmers in the traditional hog-raising areas. Ten years after the publication of "A Coming Way to Raise Hogs," 98 percent of Iowa hog producers surveyed did not have confinement-style buildings and 94 percent indicated that they had no plans to construct such facilities. For farmers who farrowed only twice per year, there was little gain to be had by investing in expensive facilities. One survey respondent stated, "Never would I go to controlled environment buildings. I raise over 500 head of hogs in the field and it's simple and easy."[70]

The minority who ventured into confinement found merit in the system. One Iowa farmer claimed in 1964 that his pigs in confinement lived "better than half the world" thanks to ample feed and fresh water,

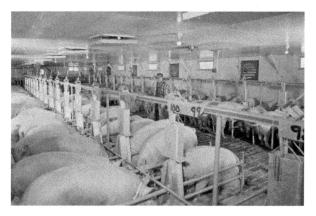

**Figure 48**

Sows chained in gestation house, circa 1966. Defenders of such practices emphasized the ease of veterinary care and the ability to provide a customized diet, while detractors attacked the practice because the animals were unable to turn around or exercise and the restraining system caused sores and abrasions. Source: Ag Illustrated. Used with permission.

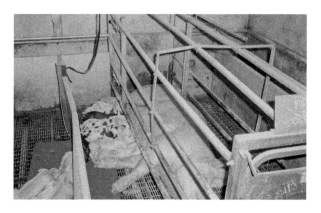

**Figure 49**

Lactating sow and pigs in a stall on the Russ Jeckel farm, 1990s. Source: Audio-Video Barn, Illinois State Museum. Used with permission.

**Figure 50**

Interior view of a farrowing barn on the Russ Jeckel farm, 1990s. Source: Audio-Video Barn, Illinois State Museum. Used with permission.

medications, and a roof over their heads.[71] Furthermore, raising pigs indoors facilitated winter farrowing, maximizing return on his investment in buildings. In 1965, a journalist described Robert Hamilton's farm near

Iowa Falls, Iowa. "The entire herd of hogs," he noted, "is kept *inside year round*."⁷²

To improve management of sows during gestation and farrowing, farmers confined them either by chain or in small stalls or crates that measured approximately six and a half by two feet. Farmers found it easier to inspect the condition of the sow, administer medication, ensure that each received the proper ration, and prevent fighting during pregnancy, thereby improving sow health and minimizing risk to fetuses. At farrowing time, experts advocated confinement in another special stall called a farrowing crate that included a space approximately eighteen inches wide on either side for the newborn pigs. The pigs could nurse, but the sow would not be able to lie on the newborns. These enclosures became standard in confinement buildings of the 1970s. Robert Hamilton explained that confinement improved the weight and strength of newborn pigs. He claimed that he often weaned ten-pig litters with rare outbreaks of scours, the old nemesis of unweaned pigs, thanks to his new farrowing house.⁷³

Damon Catron's vision of life cycle housing in segregated structures (gestation, farrowing, nursery, and finishing) combined with scientific nutrition and breeding was realized during the late twentieth century. Of course, some farmers continued to raise hogs on pasture, and experts claimed that there was still a role for traditional practices such as hogging down corn.⁷⁴ Nevertheless, rising labor and land costs, new automated material handling techniques, and the potential to apply new scientific knowledge over larger herds were all important reasons for adopting confinement-style production. The authors of a 1972 University of Illinois Cooperative Extension Service publication on confinement also noted that while "operator comfort and convenience" were difficult to quantify from an economic perspective, "it may well be the most important consideration" in facilitating the transition to new hog-raising practice.⁷⁵

When farmers moved their hogs indoors, either as part of a fully realized confinement system or in open-sided sheds on concrete pads, waste management became a major problem. It had been relatively simple to manage manure in a small hog house or pasture. A hog at or near market weight (200–250 pounds) produces as much as four times the amount of excreta as an adult human. Hogs at 50 pounds produce approximately 4.2 pounds of manure and a little over half a gallon of urine per day. A 200-pound hog, by contrast, produces 17 pounds of manure and 3 gallons

**Figure 51**
Interior view of a finishing barn on the Russ Jeckel farm, 1990s. Source: Audio-Video Barn, Illinois State Museum. Used with permission.

of urine per day. A facility with 2,000 hogs at or near market weight produced close to 15,000 pounds of solid waste and over 4,000 gallons of liquid waste per day. These figures do not include the water used to flush and clean buildings, which added to the waste. Confinement inundated farmers with manure.

To manage the manure problem, farmers used pits under the houses and large earthen structures, or holding ponds, which were called lagoons, throughout the 1990s. Pits constructed under buildings held waste prior to pumping it out and spreading it as fertilizer. The earthen holding pond contained the manure and facilitated the breakdown of waste through aerobic or anaerobic action, converting the organic matter in the waste to methane and carbon dioxide. Lagoons promised to solve manure handling problems.

Earthen structures created other problems, even as they solved problems. Farmers who built earthen structures were operating without a road map. In many cases they failed to understand the limits of these structures and provided little to no management or maintenance. Earthen structures leaked, overloaded, overflowed, breached, and always smelled horrible. According to a 1969 Purdue University Extension Service publication, "the public's demand for the control of pollution cannot and will not be denied, and the swine industry cannot revert to past systems of production to avoid the problem." Swine producers, the authors counseled, must do "a more acceptable job of disposal in the future," presumably refining and improving on the first generation of lagoons that did not function as planned.[76]

Farm experts designed new earthen structures to avoid some of the failings of the 1960s and 1970s, coaching farmers on their proper use. Purdue's agricultural engineers reminded farmers of the dangers of dis-

**Figure 52**
Exterior view of a finishing barn and earthen manure structure.
Source: Angie Rieck-Hinz, Iowa State University Extension.
Used with permission.

charge or breach of lagoons and advised them to build lagoons that would have enough capacity to handle the farm waste but also accommodate the average precipitation and major rain events that would fall on the lagoon and in the associated upland. In 1995, highly publicized manure leaks at Premium Standard Farms in the rolling hills of north-central Missouri prompted the construction of secondary containment structures to prevent leaks from reaching surface water.[77] All lagoons built in Indiana after 2002 were required to accommodate 180 days of storage capacity and at least two feet of freeboard to allow for major rainstorms without overflowing.[78]

Since the late 1990s, however, there has been a trend away from earthen structures toward the use of pits underneath the confinement building. In 1998, USDA's Economic Research Service indicated that 55 percent of hogs lived on farms with a lagoon system; 37 percent on farms with a pit system. In 2009, by contrast, the respective values were 34 percent and 62 percent.[79]

The physical landscape of hog raising saw further changes with the development of multisite production. Multisite production involves a distinct physical location for each phase of production (breeding and farrowing, nursery, and finishing), separated by a minimum of two miles. It was rooted in an earlier practice of isolated weaning (isowean), which farmers and researchers found reduced disease losses. In 1988, Sand Livestock built the first multisite production facility near Columbus, Nebraska. Numerous multisite facilities followed. Experts claimed that during the 1990s it was a critical tool in limiting losses to pseudorabies, a disease spread through nasal secretions. In infected herds, pseudorabies resulted in near 100 percent mortality in pigs younger than one month.[80] Multisite production promised to minimize disease risk at all stages of production. One of the aspects that made multisite production successful was a practice called all-in/all-out (AIAO). It was a technique borrowed from the poultry industry to break the disease cycle of reinfection. In AIAO, pigs from each farrowing facility stay together from the nursery to the finishing barn. By contrast, in continuous flow facilities, pigs of different sizes and ages and from different farms were mixed, exposing those animals to new pathogens and increasing risk of disease loss.[81]

Breeding remained a critical issue on the hog farm. One improved boar could impregnate a limited number of sows and gilts. During the 1950s, the dairy industry had success with artificial insemination (AI), using genetics to boost milk production across the herd. Hog farmers hoped for a similar application. There was considerable buzz about AI in the early 1960s, especially relating to the desire on the part of farmers and packers to create the meat-type hog.[82] Chemical estrus control helped spread the practice. By the late 1970s, however, many farmers abandoned AI. The biggest complaint was that conception rates were lower when compared to traditional breeding. In 1978, one Minnesota farmer with seven hundred sows reported that his conception rate the previous year was a dismal 48 percent, despite regular checks to see if the sows and gilts were in estrus.

AI was resurgent in the 1990s. Much of this increase was due to improvements in synchronizing estrus through the practice of simultaneous group weaning and the administration of synthetic hormones. Improved record keeping and enhanced understanding of genetics reduced the barriers to acceptance. As an Iowa State University swine geneticist claimed in 1994, "AI will take over this country in a big way."[83] Using the expertise from the Netherlands, US farmers began to establish the labs necessary for AI to spread. By 2000, approximately 70 percent of all sows in the United States were bred using AI, up from just 8 percent in 1991.[84]

Thanks to new breeding techniques, antibiotics, and the technology of confinement, farmers rapidly abandoned the pasture system and replaced it with confinement. Supporters claimed that efficiencies through economies of scale and the enhanced ability to market hogs year-round improved the lives of farmers and consumers. It was an industry triumph, but with new techniques came new criticism. Opponents claimed that CAFOs constituted a violation of property rights due to diminished air quality and pollution, while defenders claimed that oversensitive neighbors could not limit their property rights to use land as they wanted.[85] Many critics considered the rise of CAFOs as part of a race to the bottom that sacrificed animal health, the environment, as well as workers and the communities that depended on industrialized hog production, all for the sake of profit.

## THE LIMITS OF CONFINEMENT

Public criticism of hog farming was a new development of the late twentieth century. Few people outside of farm circles ever discussed pig production, let alone criticized it. Furthermore, hogs were commonplace across much of rural America, and the role they played as "mortgage lifters" made it easy for rural people to tolerate localized offending odors. For these people, complaints about the smell of the hog lot could be either dismissed or accepted with the long-standing joke "smells like money."

By the end of the century, the consensus that hogs smelled like money was under siege, even in rural America. In 1978, the Pork Motel, a four-thousand-head hog finishing operation located in central Kansas near Salina, ceased operation due to odor complaints. After detailing the efforts to control the smell, manager and part owner Bill Bowen advised future feeders to "be real careful where you build it [the lot]" and understand each season's prevailing winds. He also urged feeders to be considerate of neighbors. "Take their feelings into account," Bowen stated, "and look at problems from their point of view."[86]

As Bowen's comments suggested, most complaints about smell did not originate with urbanites or even urban refugees who sought a rural idyll. Country people, many of them hog farmers or former hog farmers, endured the worst of it. In a letter to the editor of *Successful Farming* in 2003, an Iowa farm owner from the Webster City area who was expanding his operation bluntly stated, "Hogs do stink. The industry needs to keep that in mind when they put 4,000 head in such a small space."[87] Mabel Bernard lived on her Dundy County, Nebraska, farm since 1926, but things changed in 1998, when a thirty-six-thousand-hog CAFO opened less than a mile from her home. Bernard conceded that the smell was not constant, but it was nauseating when the wind turned, causing her eyes to burn and even waking her up at night. "I can't plant enough flowers to drown it out," she explained.[88] Barbara Philipp of Ochiltree County, Texas, complained of the smell from a hog operation less than a mile from her home. "You get that odor in your house, and you can't get rid of it," she claimed, "and you cannot stay outside where it is."[89]

While the smell by itself can be awful, the effects of hog odor are more severe. Researchers have documented how people who lived near moderate to large-scale hog farms reported higher rates of tension, anger, fatigue, confusion, and diminished vigor compared to those in control groups.[90] Skeptics dismissed mood change due to subjectivity of sensation and the self-reporting of results, but ammonia and hydrogen sulfide emissions cause and contribute to asthma, bronchitis, and rhinitis, while other effects include nausea, headaches, and even diarrhea.[91] According to one study, workers in swine confinements breathe dust consisting of "swine skin cells, feces, feed, bacteria, and fungi," not to mention the many gasses that were present due to large quantities of urine and feces.[92] Of course, diminished air quality is not simply a risk for people. Even in 1971, studies indicated that hogs in confinement showed symptoms of irritation from ammonia and hydrogen sulfide.[93]

Thanks to odor concerns, neighbors of newly opened confinement farms learned that their property values declined. In 1996, economists in North Carolina concluded that housing prices in an area with high hog densities fell by 9.5 percent. Home values declined 3.6 percent where hog operations opened between a mile and a half and two miles away. A study of five counties in north-central Iowa published in 2005 indicated that the property value of a residence located downwind of a three-thousand-head hog-feeding operation declined 9 percent.[94] Such a loss was devastating in a country that often reckoned retirement savings in home values.

The experience of Susan and James McKnight was an extreme case of property devaluation. The McKnights lived in rural Sac County, Iowa, since the late 1970s, but in 1997 Iowa Select Farms built a thirty-one-thousand-head hog farm across the road. They filed a nuisance suit and won in 2002. Their home, however, was not salable. In 2003, the McKnights hired a crew to dig a hole, knock the house into it, and burn it. A writer for the *Des Moines Register* highlighted that the McKnights had lived in peace with the hog business for twenty years, but they could not accommodate the new regime. "They were used to hog odor," the *Register* explained, "but not the kind they smelled from the Iowa Select Farms facility."[95]

In addition to air pollution and odor, critics have highlighted surface and groundwater contamination. In North Carolina, heavy rainfall and massive flooding exposed the weakness of waste lagoons in 1995. Most of the new-style operations were located in the state's eastern coastal plain, and twenty-two million gallons of waste entered the New River drainage system when several structures were breached. In the aftermath of the New River incident, inspectors visited other farms and found that 124 lagoons were at capacity and 526 were overloaded. According to *U.S. News & World Report*, the overloading of North Carolina's rivers with nitrogen-rich manure killed an estimated ten million fish, resulted in the closure of 364,000 acres of coastline to shellfish harvesting, and put hundreds of commercial anglers and river guides out of business during peak season.[96]

What about water for human consumption? Several scientific studies have confirmed the old adage that shit flows downhill. Scientists compared water quality downstream and upstream of swine confinements and learned that far more fecal matter was present downstream. This was not just a case of manure in the water, but the fact that antibiotic-resistant bacteria such as enterococci and fecal coliforms were also present. While scientists noted that multiple sites could contribute antibiotic-resistant bacteria, they concluded that leakage from manure pits or runoff from hog manure applied to fields was the most likely source of contamination.[97] Earthen manure structures also collected and concentrated the growth-promoting antibiotics that passed through the animals. A US Geological Survey and Environmental Protection Agency study found that wells close to hog lagoons tested positive for the antibiotic tetracycline, commonly used in hog production.[98]

Water quality concerns galvanized Nebraskans in the communities of Orleans and Alma. In the 1990s, citizens opposed a planned thirty-four-thousand-head hog operation located five and a half miles from Alma's city wells and three miles from the site where Orleans planned to sink new wells. After the Nebraska Department of Environmental Quality authorized construction permits, community leaders hired engineers to assess the environmental impact on the local water supply. That study concluded that the combination of local porous soil used to construct earthen manure containment structures and the center pivot sprinklers used to spread the liquid manure would result in polluted groundwater and contaminated city wells within three years.[99]

In North Carolina, a disproportionate share of CAFOs were in low-income and heavily minority areas. Studies indicated that there were nineteen times more hog farms in the most impoverished areas than in the wealthiest areas and seven times more hog farms in minority areas than in mostly white areas. Many residents of eastern North Carolina relied on well water rather than municipal water, exposing them to greater risk for contamination. One study even showed that middle schools with a higher percentage of poor students tended to be closer to CAFOs.[100] Producers passed the noneconomic costs on to those who could not or would not leave.

Even so, promoters insisted that industrialized hog farms were a positive economic development, although the extent to which local communities shared in the wealth generated on these farms is arguable. Given the economic disruption of postwar globalization, communities readily sought or welcomed large-scale hog farms. Local leaders in Bladen County, North Carolina, welcomed the hog industry after four major manufacturing plants closed within a year. As the mayor of Rocky Mount explained in 1996, "It's a trade off. Jobs, tax base, [and] economic growth versus possible environmental negatives."[101] Yet in many communities, local authorities or state officials who aggressively encouraged large-scale hog farming with tax incentives were slow to punish bad actors. State and local taxpayers funded the expansion of Seaboard into Oklahoma with tens of millions of dollars, even after Albert Lea, Minnesota, had invested its own millions in keeping Seaboard there. Seaboard appealed the tax assessments subsequently enacted by Guymon, Oklahoma, and won on the basis that it created manufacturing jobs and was, therefore, exempt from taxes. Oklahoma settled a number of lawsuits against Seaboard with modest cleanup fines of approximately forty-five thousand dollars and promises of improved environmental controls. Seaboard paid no damages to neighbors.[102]

There were other costs, too. Retail spending in Illinois communities with large-scale hog operations declined compared to that in communities without CAFOs. A study of one Oklahoma community indicated that three years after the establishment of a CAFO, per capita wages decreased significantly at the same time as wages across the state increased. The county stretched services to meet the needs of low-wage migrant workers. The result, according to the North Central Regional Center for Rural Development, was that "schools became burdened, poverty increased, school dropout rates soared, and crime skyrocketed."[103] In Guymon, Oklahoma, Seaboard paid no school taxes even as the number of students in the district increased by over three hundred due to Seaboard employees, shifting the growing tax burden onto local citizens.

When asked about the economic benefits of big hog in the early 1990s, a group of North Carolinians gathered at a rural church broke into laughter. "Yeah, we heard that too," one Tar Heel explained. "But there ain't no jobs."[104] In 2004, long after the expansion of large-scale hog farming in Iowa, rural residents ranked "prisons, solid waste landfills, slaughter plants, and sewage treatment plants" over hog confinements as preferred forms of rural development.[105] Given the net effect of degraded air quality near large-scale hog farms and contamination of water and land through excess nitrates, antibiotic residue, and *E. coli*, critics were justified in wondering whether it was all worth the promised economic benefits.

Workers have borne many of the industry's hazards. The US Occupational Safety and Health Administration reported that meatpacking workers confronted two and a half times the rate of injury and illness compared to the national average, with injuries requiring time off or workplace restrictions occurring at three times the national average. As one labor lawyer in North Carolina noted, cases from Smithfield's Tar Heel plant included "a lot of back injuries, a lot of machinery injuries, fingers, hands smashed, stab and cut wounds." For nonunion workers, those risks were even greater. At the Tar Heel plant, a nonunion facility, workers who complained about injuries were simply fired.[106]

There has been significant resistance to the expansion of large-scale hog farms. In North Carolina, it began when thirty million gallons of manure washed out after record rainfall in 1995. A state legislator explained, "Some people are saying the General Assembly can't be held responsible for an Act of God. Well, I say maybe this is God telling us it's time to do something before things get worse."[107] In 1996, overflowing North Carolina manure structures in the coastal plain due to Hurricanes Bertha and Fran provoked renewed concern and protest about the unregulated industry labeled "Boss Hog" in a Pulitzer Prize–winning series of exposés by two *Raleigh News and Observer* journalists. In 1997, the North Carolina legislature passed the Clean Water Responsibility Act, which mandated a two-year moratorium on expansion of hog farms and the construction of any new facility for more than two hundred fifty animals. Meanwhile, Hurricane Floyd washed an estimated 120 million gallons of hog waste into eastern North Carolina's estuaries and rivers. In 1999, the legislature renewed the moratorium for all facilities that used lagoons for waste but exempted those that used "environmentally superior technologies."[108]

In Missouri, legislators allowed an exemption to the state's anti-corporate farm legislation, granting Premium Standard Farms permission to operate in three counties in the north-central part of the state. In 1994, Daviess County became home to twenty thousand sows and their half a million pigs. Martha Stevens, a farmer and activist from Hatfield, pointed out that Continental Grain's hog farms near Bethany produced more sewage than Kansas City, while Premium Standard Farms hog farms located in northern Missouri generated five times more waste per day than Kansas City. Neither company was required to treat their waste, she noted, yet small towns were legally required to have waste treatment facilities. "I can't see how a few people are a health hazard and two million hogs aren't [hazardous]," Stevens observed.[109] New legislation in 1996 provided for the regulation of Missouri's CAFOs, although it applied only to future operations, and then only the largest ones.

Concerned residents of the Rosebud Sioux Reservation in South Dakota complained when tribal leaders and the Bureau of Indian Affairs (BIA) approved a lease for Bell Farms to create a four-thousand-acre facility for 859,000 hogs on the reservation, complete with five hundred fifty acres of manure lagoons. The lure of economic development was attractive for tribal leaders. Unemployment on the reservation was high when Bell proposed creating two hundred jobs on the reservation and returning 25 percent of the profits to the tribe. As former tribal president Normal Wilson stated, "I want to put the head of the family back to work and providing for his family."[110]

Not everyone on the reservation agreed. Eva Iyotte and Oleta Mednansky recognized that there had not been an environmental impact statement prepared prior to the granting of the lease and construction, as required by the National Environmental Policy Act. They organized as Concerned Rosebud Area Citizens (CRAC) and sued the BIA in 1998, voting out the tribal leaders who authorized the lease. The Justice Department rescinded the lease, but Bell Farms filed to overturn the settlement. A federal judge agreed with the company and construction of the first phase continued, but a federal appeals court supported CRAC's claim and the US Supreme Court refused to hear the case on appeal. It was a victory for the citizens' group, although the company continued to operate a ninety-six-thousand-head facility that had already opened.[111]

Such setbacks did not constitute defeat for intensive hog farming. Defenders of the large-scale system argued that expanded production produced economies of scale that were necessary to sustain American agriculture. As the chief financial officer of Sand Livestock wrote in his testimony to the Nebraska legislature in 1997, opponents of large-scale farming had a nostalgic vision of American agriculture. "Their agenda," he claimed, "is to keep the farmers small, and rural America poor and dependent on government handouts. Frankly, these people seem to be on a mission to destroy anyone who can demonstrate that American farmers are capable of being more than peasants or serfs."[112] This characterization of the opposition to CAFOs, however, failed to engage the real problems that neighbors witnessed and endured.

The problem for both communities and the industry was that in the mid-1990s much of the evidence about hazards was anecdotal. To rectify this, scientists gathered in Des Moines to deal with public concerns and complaints, independent of pork industry support or sponsorship and in the midst of Iowa's expansion of large-scale hog operations. In October 1994, they held a listening forum to collect questions from the public and farmers and convened a scientific workshop the next June to establish the most reliable answers to those questions. As two of the planners noted, scientists were in a bind. Contending constituencies too often asked them to simply justify existing positions rather than present the most accurate and complete information. In 1996, they issued a final report that indicated how little was known about the new regime of confinements.[113]

Early in the section on water use, scientists noted that they were unaware of any "systematic study of actual water use by [confinement] facilities." They also conceded that there were no known methods of detecting microbial contamination of water due to swine-specific sources. Similarly, there had been no field survey of manure lagoons to determine their working life span and no comparative study of first-generation lagoons with newer models to assess just how well they worked.[114]

By contrast, scientists knew more about the health of swine-industry workers. The largest number of complaints and the best-known issues were respiratory ailments, most commonly bronchitis. Other common ailments included sinusitis, rhinitis, occupational asthma, and organic toxic dust syndrome. The author labeled swine workers as a "survivor population," which simply meant that people who were most severely affected by their work in the industry quit and were not included in the data. Those who remained, however, suffered from chronic conditions due to employment in the confinement barns. In any case, it was possible to minimize risks for workers by monitoring (health surveillance) and management (proper equipment), both of which were "exceedingly rare in today's [1995] swine industry."[115]

The report also revealed the contested nature of economic benefits. The author of this section suggested that there were really two frequently conflated issues at stake, economic growth and economic development. Growth tended to be more quantifiable, with jobs, payroll, and prices as key indicators. By contrast, development was qualitative, indicated by education, health, and community investment. Large-scale hog farming tended to score well in terms of economic growth but less so in terms of development. The author suggested that it was possible even to conceive of economic growth "to the extent that it results in thwarting economic development."[116] If growth increased while development stagnated, communities that relied on hog farming were stuck with some short-term benefits and long-term liabilities.

Industry expansion continued in the 1990s, but there was push back. The most successful opposition occurred over issues of air quality rather than land and water contamination. In 2010, a Missouri court awarded eleven million dollars in damages to fifteen plaintiffs in an odor case. The defendants, Premium Standard Farms, Inc. (a subsidiary of Smithfield Foods) and ContiGroup Companies (formerly Continental Grain), were to pay for eleven years of damages. In 2012, a different group of three hundred Missourians who sued PSF won another settlement.[117]

The attorneys who argued for the plaintiffs, Richard Middleton and Charlie Speer, have secured thirty-two million dollars in odor settlement jury awards in just ten years. They prevailed in large part because they focused on the only aspect of large-scale CAFOs that they could win—odor. Not surprisingly, industry fought back by shaping state legislation to limit lawsuits and cap damages. Yet as one of the victorious attorneys explained, there is very little that neighbors and downstream communities can do about pollution. The nuisance suits have given some people hope. As of early 2014, Middleton and Speer had filed suits on behalf of four thousand North Carolinians and over five hundred Iowans. "We can't shame them and we can't reason with them," attorney Charlie Speer observed. "The only way to get their attention is to go after their profits."[118] An industry that transformed itself to maximize profits is less sensitive to commonplace concerns over environmental contamination and more sensitive to anything that would compromise the bottom line.

The bottom line, the principal reason invoked for the widespread intervention of science and technology in the hog business, is itself threatened by those scientific and technological solutions. As indicated in chapter 6, hogs in the new mega-confinement systems were hyper-vulnerable to pathogens. The scale of production meant that any disease outbreak is difficult to isolate. Despite the prevalence of AIAO, the disease problem in modern hog production has not ceased, as shown most recently by the PRRS and PEDv outbreaks.

The widespread use of subtherapeutic antibiotics that facilitated rapid growth and reduced mortality rates among newborn pigs proved to be a double-edged sword. The downside is that antibiotics lose their effectiveness. The prospect of antibiotic resistance was a concern as early as the 1970s, but industry groups successfully fought against proposed restrictions by the FDA. In December 2016, *Scientific American* published an essay that noted how recent studies showed that antibiotic resistance in

**Figure 53**

Watering hogs on the Biesecken farm, near Frankfort, Indiana, in 1965. Each of these lots was home for 150 hogs during the summer. In recent years, a significant number of hog farmers have returned to pasture systems, although the overwhelming majority of swine are raised in CAFOs. Source: Ag Illustrated. Used with permission.

livestock passes to offspring and can be shared across species boundaries. The reality is that the more we learn about resistance mutations, the more we realize just how little we know. The knowledge vacuum, however, is not an asset for those who want to limit antibiotic use. There is little motivation for individual producers to restrict or abandon antibiotics, even as resistance threatens to become a larger problem not only for livestock producers but for everyone who relies on antibiotics.[119]

In at least one sense, the old has become new. For a minority of producers, bigger is not necessarily better. The same *Scientific American* article that highlighted the threat of superbugs due to agricultural production featured a farm in Indiana that had rejected antibiotics as a regular part of the ration. Mangers at Seven Sons Farms weaned pigs later and raised animals on pasture in lower density than in confinement structures. They relied on the animal's immune system rather than on regularly administered antibiotics to do more of the work in controlling losses.

The rejection of confinement by a handful of farmers such as Niman Ranch and Seven Sons Farms did not entail a rejection of science and technology. Instead, it was an acceptance of some of the limits of science and technology. Americans confronted limits before, but much of the public awareness of those limits occurred because these problems associated with limits were visible, often with an urban component. As long as hog production continues to occur out of sight in areas of low population, it will be difficult for Americans to accept the fact that science can best aid the swineherd when we accept the reality of scientific and environmental limits.

CODA

# THE FUTURE OF HOGS IN AMERICA

Swine were critical for success in constructing an economic, social, and political empire in America, even if Americans disputed their place in that empire. Farmers, experts, and consumers fought about what constitutes good farming, the proper role of pork in the diet, and the place of pigs in the landscape. Elected officials, elites, and have-nots argued over the various ways in which pigs occupied urban and rural places. People enlisted governments to intervene and affirm the place of hog farming in the nation through disease control efforts, environmental regulation, tax incentives, and labor laws. Thousands of human decisions challenged or supported prevailing assumptions about the place of pigs and pork over the past five hundred years.

Studying the history of hogs reveals that much of the story was a runaway success, transcending many of the traditional limits on production and consumption. Swine are present on the land in greater numbers than ever before. Pork and lard are common foods and constitute a major American export. The technological transformations in breeding, feeding, health, and housing accelerated the production process and reduced consumer costs.

There were important limits, however. Despite massive efforts to boost consumption, pork promoters failed to challenge the place of beef and chicken during the late twentieth century. Municipal leaders banished pigs from cities and the urban margins, making it difficult to use swine in waste reduction due to high transportation costs. The animal industry's failure to adequately manage the externalities and environmental impact of hog production remains a problem. Water and air quality issues are pervasive everywhere large-scale hog production occurs. Despite major investments in eradicating hog cholera, current production techniques compound the risk of epidemic disease. For all the economic merits of today's production hog, the feral hog is sturdier and a far more vital force on the land. In short, the success of pigs, pork, producers, and processors is not the whole story.

If the history of pigs in America shows that many dreams were unfulfilled, is it possible to provide some clues about the future? Pork and lard will continue to play a large role in the American diet. Lard is valued again, due to nostalgia for grandmother's cooking in an age of fast and convenient food and a recognition that our obsession for cutting animal fat led us to replace it with sugar, salt, and other kinds of fat. Pork may even play a larger role in the American diet in the coming years. The desire for salty bacon and charcuterie as retro-haute cuisine fueled the popularity of pork even in the midst of debates over the healthy qualities of pork and lard, the ethics of animal agriculture, and the sustainability of contemporary production techniques. Sustainability issues could actually strengthen the position of pork in the marketplace. Recent studies indicated that greenhouse gas emissions produced by the pork industry are low compared to beef production. In 2011, the US Environmental Protection Agency reported that the livestock business accounted for approximately 2.7 percent of the nation's total greenhouse gas emissions: pork was only one-third of 1 percent of that total.[1] Herbivorous ruminants such as cattle require far more biomass to convert the sun's energy into flesh than omnivorous hogs, resulting in a greater carbon footprint to produce a pound of beef than of pork. While other sources of protein may be less resource-intensive (think insect protein or laboratory-grown meat), the reality of animal agriculture is not likely to go away soon and pork may be one of the better meat options.

Swine may play a renewed role in waste reduction. For a renewal to occur, however, the geography of hog farming will need to shift, with animals moving back closer to population centers. Such a move risks alienating suburbanites who are concerned about odor from manure and garbage on the city's margins. But if Americans redevelop central cities and reverse urban and suburban sprawl, space on the margins may be viable for waste feeding.

Researchers continue to explore techniques to control odor and safely manage manure. The manure from CAFOs threatens the safety of ground and surface water. *E. coli*, nitrogen, antibiotic residue, and the massive amounts of manure from even a modest hog farm can contaminate local water and that of downstream communities and ecosystems far removed from the source. The increasing costs to those downstream communities for water quality mitigation may become great enough to pressure the industry to change. Industrial hog production also threatens air quality. More than odor affronts neighbors and compromises property values: noxious gasses and airborne particles threaten the health of neighbors and workers.

If the past is prologue, there will be significant changes in the productive technology of hog farming. Mapping the swine genome will result in new applications of biotechnology to the business of making animals into meat as well as managing waste. Experts are studying such problems as new diseases and marketing. The caution is that the process of reinventing production will bring changes that cut multiple ways, just like the major technical production innovations of the past.

In our relentless concern over the bottom line, wages have fallen precipitously for many of the workers who convert animals into bacon, chops, and ham. Meatpacking was once a job that provided a degree of economic security for working-class Americans. In contemporary America, undocumented workers perform much of this work, often for short terms and for wages that many native-born Americans consider too low. Current political debates over immigration, both documented and undocumented, will have serious consequences for meatpackers. Stricter immigration laws will likely result in increased wages, potentially leading to higher food costs. A victory for those who favor more liberal immigration policies will affirm the status quo in the meat industry and keep food prices comparatively low.

In the near term, however, much of the industry will resemble the status quo. Environmentalism and environmental regulation are dirty words in much of the country, especially in the major hog-raising states of the South, Midwest, and Great Plains. Given the current reluctance to raise the minimum wage in these locations, it is difficult to imagine circumstances that would result in higher wages for laborers. Furthermore, advocates for a revival of pasture-based systems or alternate techniques recognize that we have doubled down on industrialization since the 1980s. The concentration and integration that prevailed over the past generation have significant momentum, which constitutes a considerable obstacle to change. While efficiency was the motivation for increased scale, increased scale does not necessarily mean more efficiency in producing a pound of pork. Economist Mike Duffy of Iowa State University explained, "It's not about efficiency at all. It's about power."[2] Larger entities such as Smithfield have more power than any group of smaller producers and therefore have more influence in the marketplace and in determining policy.

What about the farmers who continue in the pig production business, including the large-scale enterprises, contractors, and independent producers? While it is difficult to generate much sympathy for the corporate leaders and integrators who are more concerned about shareholders and the bottom line than about communities, it is important to remember that many farmers and farm wage workers care about the animals they raise and the communities where they live. As one farm blogger wrote about her experiences on a Minnesota hog farm, "Comfortable and content pigs is what is best." She explained that many farmers were frustrated because critics could not see the ways hog farmers intervened on behalf of the animals on their farm.[3] Producers repeatedly negotiated those interventions on private and public experimental farms, in cities, and in print for hundreds of years. Farmers have been malleable in the past, adopting new techniques and forms of organization in the name of science and market conditions, and it is likely that they will be amenable to change in the future too.

The place of hogs in America is likely to get a boost from the medical community. The biological similarities between humans and hogs opened the door to the use of living tissue across species boundaries, a practice known as xenotransplantation. The first corneal transplant was from a hog to a human in 1838. In 1907, a French surgeon who pioneered the surgical connection of blood vessels predicted that xenotransplantation was possible, stating, "The ideal method would be to transplant in man organs of animals easy to secure and operate on, such as hogs, for instance."[4] In recent years, many Americans have benefited from replacement heart valves from pigs. Many more people will have more of the hog within them in the future, including entire organs rather than parts.

While hog organs for transplantation make sense given the biological similarities of swine and humans, primates were the first candidates medical researchers selected in the 1960s. The most famous xenotransplantation case was that of Baby Fae, a child born with a fatal heart defect. In 1983, Baby Fae received a baboon heart transplant and survived twenty days after surgery, attracting widespread media attention. The transplant reflected a growing sense that xenotransplantation could reduce human suffering and solve a host of human health problems. "In the future," a writer for *Newsweek* predicted in 1993, "farmers will raise animals whose organs will be transplanted into humans."[5] By the mid-1990s, an estimated three thousand Americans died every year while waiting for an organ transplant; another hundred thousand died never having qualified for a place on the transplant waiting list.[6]

For many Americans, pigs were better suited for transplantation than were primates. In addition to the previously noted human-hog biological similarities, cultural considerations favor hogs. During the late twentieth century, there was a backlash against medical experimentation and testing on primates. By contrast, most Americans were comfortable with the practice of killing hogs and consuming them as food and using their by-products. Our shared history is one of harvesting hogs to meet human needs and wants.

Why has xenotransplantation taken so long to unfold if Americans are accustomed to utilizing hogs and they are the best physiological match? Some severe technical obstacles remain unresolved. Transplantation requires suppression of the human immune system to prevent the body from attacking the new tissue, which is an even greater problem across species boundaries. A sugar molecule known as Gal covers pig blood vessels, but it is not present in humans, which means that the human immune system immediately rejects the new organ. By contrast, the pig heart valves used for the human heart are chemically treated to prevent rejection. Whole-organ transplant is much more complicated than valve replacement and has yet to be successful on a large scale due to the problem of human anti-Gal (a-Gal) antibodies.[7]

Genetic modification of the pig promises to yield organs that lack Gal and are acceptable to the human host. Pig embryos from specific-pathogen-free pigs (raised under strict conditions to prevent the presence of microbes) are surgically harvested and then a human gene is inserted. Scientists hoped for a transgenic pig that would yield an organ without provoking hyperacute rejection and would allow the widespread use of hog organs in humans.[8]

These developments did not address the more basic question of whether xenotransplantation was proper. In 2001, the *New York Times* featured side-by-side editorials debating the risks and benefits of xenotransplantation. Dr. Jonathan Allan, DVM, contended that risks of viruses migrating across the species boundary were too great. "It's one thing for a desperate patient to put himself at risk by giving consent for an experimental procedure," Allen stated, but with xenotransplantation, "an infection picked up from an animal organ could circulate into the general population." Furthermore, if these transplants were widely available, who would pay for such expensive procedures?[9] Dr. David Cooper countered that the risk of virus migration was low and that there were pigs under study that were incapable of transmitting viruses to human cells. Focusing on infection from pigs obscured the fact that human organ transplants also run the risk of infections. Cooper conceded that although the costs of xenotransplantation were high, other treatments such as dialysis and end-stage organ failure were even more expensive. Cooper also noted that many donor organs from humans were not in optimal condition at the time of transplant due to age, brain damage, or poor circulation, all problems that fresh, predictable organs from specially bred hogs could solve.

For all the hype, xenotransplantation faded from public consciousness in the early 2000s. Investment faltered due to the seemingly intractable problem of surmounting the cross-species immunity problem and the risks of introducing pig viruses into the human body. As one scientist stated, xenotransplantation became "the third rail of biotech . . . as a business plan."[10]

Yet new hope arose in 2015. Funded by United Therapeutics, researchers developed transgenic pigs with multiple pieces of human DNA. By that year, scientists inserted as many as eight human genes into swine to prevent a hyperacute immune response in the human host.[11] Using transgenic hogs bred in Blacksburg, Virginia, researchers kept a baboon alive for two and a half years with an a-Gal-free pig heart, suggesting the potential to overcome the immunity problem. In another experiment, researchers transplanted a pig kidney into a baboon, which kept the animal alive for 136 days. These timelines may seem modest to members of the public, especially those waiting for organ transplants, but xenotransplantation experts were excited about measuring survival time in months and years rather than seconds and minutes. In addition to transplanted lungs, hearts, and corneas, insulin made from hog islet cells may soon be part of our bodies.

Recent stem cell research and gene editing has promised to make xenotransplantation a reality. In early 2017, the *New York Times* reported on the possibility of using human stem cells from a patient to grow replacement organs inside the body of a pig. In theory, at least, there would be no immune rejection problem because the new organ would be composed of the patient's own cells. Utilizing gene-edited pigs as hosts to grow organs for transplant that would not provoke the immune response could resolve one of the major problems in xenotransplantation research.[12] Ever more of the hog is likely to occupy the human body in the coming years, promising to transcend old limits in the human-hog relationship as well as to bring new risks and failures.

# NOTES

## INTRODUCTION

1. Brett Mizelle made a similar claim for the ubiquity of pigs in the modern world in his book titled *Pig* (London: Reaktion Books, 2011), 7.
2. See Mizelle, *Pig*, and Mark Essig, *Lesser Beasts: A Snout-to-Tail History of the Humble Pig* (New York: Basic Books, 2015).
3. Patrick Vinton Kirch, *Feathered Gods and Fishhooks: An Introduction to Hawaiian Archaeology and Prehistory* (Honolulu: University of Hawaii Press, 1985), 2–3, 78–81, 291.
4. Lewis Cecil Gray, *History of Agriculture in the Southern United States to 1860*, 2 vols. (Washington, DC: Carnegie Institute of Washington, 1933; repr., Gloucester, MA: Peter Smith, 1958), 1:10–12, 107.
5. Charles Wayland Towne and Edward Norris Wentworth, *Pigs: From Cave to Corn Belt* (Norman: University of Oklahoma Press, 1950), 72; C. Allan Jones, *Texas Roots: Agricultural and Rural Life before the Civil War* (College Station: Texas A&M University Press, 2005), 87; William W. Dunmire, *New Mexico's Spanish Livestock Heritage: Four Centuries of Animals, Land, and People* (Albuquerque: University of New Mexico Press, 2013), 21–22.
6. Peter O. Wacker, *Land and People: A Cultural Geography of Preindustrial New Jersey: Origins and Settlement Patterns* (New Brunswick, NJ: Rutgers University Press, 1975), 32.
7. Gray, *History of Agriculture*, 1:19; Mizelle, *Pig*, 42–43. Gray and Mizelle accepted Smith's figures at face value, but Virginia DeJohn Anderson provided a more realistic appraisal of Smith's account in *Creatures of Empire: How Domestic*

*Animals Transformed Early America* (New York: Oxford University Press, 2004), 102. The estimate of litter size is from Lois Green Carr, Russell R. Menard, and Lorena S. Walsh, *Robert Cole's World: Agriculture and Society in Early Maryland* (Chapel Hill: University of North Carolina Press, 1991), 48.

8. David Steven Cohen, *The Dutch-American Farm* (New York: New York University Press, 1992), 119, 121.
9. Gray, *History of Agriculture*, 1:51, 104.
10. Harriet Ritvo, "Going Forth and Multiplying: Animal Acclimatization and Invasion," *Environmental History* 17 (2012): 404–414.

## CHAPTER 1

1. Robert R. Gradie III, "New England Indians and Colonizing Pigs," *Archives of the Papers of the Algonquian Conference* 15 (1984): 155; William Cronon, *Changes in the Land: Indians, Colonists, and the Ecology of New England* (New York: Hill & Wang, 1983), 162–163; Virginia DeJohn Anderson, *Creatures of Empire: How Domestic Animals Transformed Early America* (New York: Oxford University Press, 2004), 206–208.
2. Anderson, *Creatures of Empire*, 188–189, Cronon, *Changes in the Land*, 131; Mart A. Stewart, *"What Nature Suffers to Groe": Life, Labor, and Landscape on the Georgia Coast, 1680–1920* (Athens: University of Georgia Press, 1996), 27.
3. Anderson, *Creatures of Empire*, 190.
4. Cronon, *Changes in the Land*, 130.
5. Ivor Noël Hume, *The Virginia Adventure, Roanoke to James Towne: An Archaeological and Historical Odyssey* (New York: Knopf, 1998), 259–260; Allan Greer, "Commons and Enclosure in the Colonization of North America," *American Historical Review* 117(2) (April 2012): 383.
6. Edwin G. Burrows and Mike Wallace, *Gotham: A History of New York to 1898* (Oxford: Oxford University Press, 1999), 37–38.
7. Alan Taylor, *American Colonies* (New York: Viking, 2000), 199–203; Anderson, *Creatures of Empire*, 230–237.
8. David J. Silverman, "'We Chuse to Be Bounded': Native American Animal Husbandry in Colonial New England," *William and Mary Quarterly*, 3rd ser. 60(3) (July 2003): 514–515; Gradie, "New England Indians," 159.
9. Robert D. Newman, "The Acceptance of European Domestic Animals by the Eighteenth Century Cherokee," *Tennessee Anthropologist* 4(1) (Spring 1979): 102.
10. Anderson, *Creatures of Empire*, 212–214.

11. Anderson, *Creatures of Empire*, 218.
12. Mary Rowlandson, "A True History of the Captivity and Restoration of Mrs. Mary Rowlandson," in *Colonial American Travel Narratives*, ed. Wendy Martin (New York: Penguin, 1994), 37.
13. Silverman, "'We Chuse to Be Bounded,'" 542.
14. Cited in Malcolm Rohrbough, *The Trans-Appalachian Frontier: People, Societies, and Institutions, 1775–1850* (New York: Oxford University Press, 1978), 173.
15. "Fatting Swine," *American Agriculturalist*, October 1845, 297; John C. Hudson, *Making the Corn Belt: A Geographical History of Middle-Western Agriculture* (Bloomington: Indiana University Press, 1994), 151.
16. This account of the Pig War is drawn from Deborah Franklin, "Boar War," *Smithsonian* 36(3) (June 2005): 64–70.
17. John Solomon Otto, *Southern Agriculture during the Civil War Era, 1860–1880* (Westport, CT: Greenwood, 1994), 22.
18. Paul W. Gates, *Agriculture and the Civil War* (New York: Knopf, 1965), 11.
19. Otto, *Southern Agriculture*, 15, 36; Andrew F. Smith, *Starving the South: How the North Won the Civil War* (New York: St. Martin's, 2011), 33. In 1866, Edward Pollard (an outspoken critic of the Davis administration) noted the significance of the loss of the pork-producing region of Tennessee and Kentucky in *The Lost Cause: A New Southern History of the War of the Confederates* (New York: E.B. Treat, 1866), 481; US War Department, *The War of the Rebellion: The Official Records of the Union and Confederate Armies War of the Rebellion* (Washington, DC: Government Printing Office, 1900), ser. 4, vol. 2, 193, hereafter cited as *OR*.
20. Gates, *Agriculture and the Civil War*, 93; Chuck Veit, "Raiding Robert E. Lee's Commissary," *Naval History* 22(4) (August 2008): 52–58; Charles W. Porter, ed., *In the Devil's Dominions: A Union Soldier's Adventures in "Bushwhacker Country"* (Nevada, MO: Vernon County Historical Society, 1998), 116–117.
21. Michael G. Mahon, *The Shenandoah Valley, 1861–1865: The Destruction of the Granary of the Confederacy* (Mechanicsburg, PA: Stackpole Books, 1999), 123–124.
22. Alexander G. Downing, *Downing's Civil War Diary*. ed. Olynthus B. Clark (Des Moines: Historical Department of Iowa, 1916), 184.
23. Gates, *Agriculture and the Civil War*, 176–177, 183–184; Allan G. Bogue, "Twenty Years of an Iowa Farm Business, 1860–1880," *Annals of Iowa* 35(8) (Spring 1961): 567.

24. Theodore Saloutos, "Southern Agriculture and the Problems of Readjustment: 1865–1877," *Agricultural History* 30(2) (April 1956): 61.
25. Grady McWhiney, Warren O. Moore, Jr., and Robert F. Pace, eds., *"Fear God and Walk Humbly": The Agricultural Journal of James Mallory* (Tuscaloosa: University of Alabama Press, 1997), 252.
26. Tom Moore Craig, ed., *Upcountry South Carolina Goes to War: Letters of the Anderson, Moore, and Brockman Families, 1853–1865* (Columbia: University of South Carolina Press, 2011), 167.
27. Otto, *Southern Agriculture*, 100.
28. Thomas P. Janes, *A Manual on the Hog*, Circular No. 40 (Atlanta: Jas. P. Harrison, 1877), 2–3.
29. Cited in Harry J. Carman, "English Views of Middle Western Agriculture, 1850–1870," *Agricultural History* 8(1) (January 1934): 18.
30. Allan G. Bogue, *From Prairie to Corn Belt: Farming on the Illinois and Iowa Prairies in the Nineteenth Century* (Chicago: University of Chicago Press, 1963), 112–113.
31. Brian W. Beltman, *Dutch Farmer in the Missouri Valley: The Life and Letters of Ulbe Eringa, 1866–1950* (Urbana: University of Illinois Press, 1996), 80.
32. Cited in Mildred Throne, "Southern Iowa Agriculture, 1865–1870," *Iowa Journal of History* 50 (July 1952): 215–216.
33. Robert C. Ostergren, *A Community Transplanted: The Trans-Atlantic Experience of a Swedish Immigrant Settlement in the Upper Middle West, 1835–1915* (Madison: University of Wisconsin Press, 1988), 200–201.
34. *U.S. Census of Agriculture, 1930*, vol. 1, pt. 1 (Washington, DC: Government Printing Office, 1932), 246; *U.S. Census of Agriculture 1940*, vol. 3 (Washington, DC: Government Printing Office, 1943), 636; Dennis S. Nordin and Roy V. Scott, *From Prairie Farmer to Entrepreneur: The Transformation of Midwestern Agriculture* (Bloomington: Indiana University Press, 2005), 74.
35. Marvin Hayenga, V. James Rhodes, Jon A. Brandt, and Ronald E. Deiter, *The U.S. Pork Sector: Changing Structure and Organization* (Ames: Iowa State University Press, 1985), 21; V. James Rhodes, "The Industrialization of Hog Production," *Review of Agricultural Economics* 17(1995): 107.
36. Nigel Key and William McBride, "The Changing Economics of U.S. Hog Production" (USDA-ERS Economic Research Report 52; Washington, DC: USDA and ERS, December, 2007), iii.
37. Joe Vansickle, "Hoosier State Embraces Hog Growth,"

*National Hog Farmer*, May 15, 2006, http://nationalhogfarmer.com/mag/farming_hoosier_state_embraces.

38. Informa Economics, *The Changing US Pork Industry and Implications for Future Growth* (Washington, DC, 2005), 11.
39. Mark S. Honeyman and Michael D. Duffy, "Iowa's Changing Swine Industry" (Animal Industry Report: AS 652, ASL R2158, Iowa State University, 2006), np.
40. Hayenga et al., *U.S. Pork Sector*, 21.
41. John D. Lawrence, "The State of Iowa's Pork Industry—Dollars and Scents" (Iowa State University Agriculture and Home Economics Experiment Station, No. Pm-1746, January 1998), chap. 1.
42. Rhodes, "Industrialization of Hog Production," 110.
43. Lawrence, "State of Iowa's Pork Industry," chap. 1.
44. Key and McBride, "Changing Economics of U.S. Hog Production," 8.
45. Michael D. Thompson, "This Little Piggy Went to Market: The Commercialization of Hog Production in Eastern North Carolina from William Shay to Wendell Murphy," *Agricultural History* 74(2) (Spring 2000): 575–579.
46. Daniel Roth, "The Ray Kroc of Pigsties," *Forbes*, October 13, 1997.
47. Owen J. Furuseth, "Restructuring of Hog Farming in North Carolina: Explosion and Implosion," *Professional Geographer* 49(4) (1997): 395, 397–402.
48. Chris Hurt and Kelly Zering, "Hog Production Booms in North Carolina: Why There? Why Now?," *Purdue Agricultural Economics Report*, August 1993, 11–13.
49. Gene Johnston, "Southern-Style Contract Hogs Creep toward Corn Belt," *Successful Farming*, September 1986, H2–H3.
50. Dean Houghton and Karen McMahon, "Here Comes the Corporate Sow," *Farm Journal*, October 1984, 15.
51. "What Farmers Are Saying," *Successful Farming*, Mid-March 1991, 10.
52. Patrick Luby, "The Hog-Pork Industry Woes of 1998" (Marketing and Policy Briefing Paper No. 67, Department of Agricultural and Applied Economics, College of Agriculture and Life Sciences, University of Wisconsin–Madison, Cooperative Extension, April 1999).
53. Alessandro Bonanno and Douglas H. Constance, "Corporations and the State in the Global Era: The Case of Seaboard Farms and Texas," *Rural Sociology* 71(1) (2006): 67–70; Richard Lowitt, *American Outback: The Oklahoma Panhandle in the Twentieth Century* (Lubbock: Texas Tech University Press, 2006), 100–103; John Fraser Hart and Chris Mayda, "Pork Palaces on the Panhandle,"

*Geographical Review* 87(3) (July 1997): 397–399; Alex Blanchette, "Herding Species: Biosecurity, Posthuman Labor, and the American Industrial Pig," *Cultural Anthropology* 30(4) (2015): 652–653.

54. Dean Houghton and Karen McMahon, "The $50 Million Hog Farm," *Farm Journal*, October 1984, 13–15.
55. Alan Barkema and Michael L. Cook, "The Changing U.S. Pork Industry: A Dilemma for Public Policy," *Federal Reserve Bank of Kansas City Economic Review*, Second Quarter 1993, 57.
56. Bonanno and Constance, "Corporations and the State," 67–70.
57. Betsy Freese, "It's Too Little, Too Late to Alter the Course of the Pork Industry," *Successful Farming*, November 1999, 9.
58. Marvin L. Hayenga, "Cutting the Verticals Down to Size: Congress, the Farm Bill, and Packer Control," *Choices*, Summer 2002, 38.
59. Christopher Leonard, "Red Meat: Smithfield Foods and Communist China Are a Match Made in Heaven," *Slate*, June 10, 2013, http://www.slate.com/articles/news_and_politics/food/2013/06/shuanghui_s_smithfield_takeover_what_the_planned_buyout_means_for_american.html.
60. Quoted in Richard P. Horwitz, *Hog Ties: Pigs, Manure, and Mortality in American Culture* (New York: St. Martin's, 1998), 234.
61. "Business: Slaughterhouse Rules: Competition in American Agriculture," *Economist*, June 26, 2010, 68.
62. R. P. Hanson and Lars Karstad, "Feral Swine in the Southeastern United States," *Journal of Wildlife Management* 23(1) (January 1959): 64–74.
63. John J. Mayer and I. Lehr Brisbin, Jr., *Wild Pigs in the United States: Their History, Comparative Morphology, and Current Status* (1991; Athens: University of Georgia Press, 2008), 203; Amy Nordrum, "Can Wild Pigs Ravaging the U.S. Be Stopped?," *Scientific American*, October 21, 2014, http://www.scientificamerican.com/article/can-wild-pigs-ravaging-the-u-s-be-stopped/.
64. Chris Jaworowski, "Feral Hogs-Wildlife Enemy Number One," *Outdoor Alabama*, http://www.outdooralabama.com/feral-hogs-wildlife-enemy-number-one.
65. Piseth Tep and Katrina Gaines, "Reversing the Impacts of Feral Pig on the Hawaiian Tropical Rainforest Ecosystem," *Restoration and Reclamation Review* 8(3) (Fall 2003): 1–3; Jack E. Rosenberger, "Attack of the Feral Pigs," *Environmental Magazine* 5(5) (September/October 1994): 22.
66. Brian T. Murray, "Gone Hog Wild! Gloucester Bristles as Feral Porkers Run Amok," *Newark Star-Ledger*, August 23, 2008.

67. Joseph L. Corn, James C. Cumbee, Brian A. Chandler, David E. Stallknecht, and John R. Fischer, "Implications of Feral Swine Expansion: Expansion of Feral Swine in the United States and Potential Implications for Domestic Swine," in *Proceedings of the One Hundred and Ninth Annual Meeting of the United States Animal Health Association* (November 3–10, 2005), 295.
68. Nordrum, "Can Wild Pigs Ravaging the U.S. Be Stopped?"; Bob Butz, "Fat Chance," *Outdoor Life* 214(1) (December 2006/January 2007): 100–104.
69. Chad Mason, "Hogs the Hard Way," *Outdoor Life* 213(7) (August 2006): HB 10–HB 13.
70. Jean Garner, "Wild Pigs Menace U.S., Whet Appetites in Europe," *ABC News*, December 12, 2006, http://abcnews.go.com/Nightline/US/story?id=2717845; Jackson Landers, "Want to Help the Environment? Go Shoot a Pig," *Slate*, August 9, 2012, http://www.slate.com/articles/health_and_science/science/2012/08/hunt_wild_pigs_for_the_environment_kill_and_eat_invasive_species_.html.
71. Landers, "Want to Help the Environment?"; Ben C. West, Andrea L. Cooper, and James B. Armstrong, *Managing Wild Pigs: A Technical Guide. Human-Wildlife Interactions Monograph 1* (Starkville, MS and Logan, UT: Berryman Institute, 2009), 33; Mississippi State University Extension, "A Pickup Load of Pigs: The Feral Swine Pandemic," *Wild Pig Info*, http://wildpiginfo.msstate.edu/index.html.
72. West, Cooper, and Armstrong, *Managing Wild Pigs*, 24–38.
73. Avi Selk, "'Hog Apocalypse': Texas Has a New Weapon in Its War on Feral Pigs: It's Not Pretty," *Washington Post*, February 23, 2017; Ryan Kocian, "Texas 'Hog Apocalypse' Poison Put on Hold," *Courthouse News Service*, March 7, 2017, https://www.courthousenews.com/texas-hog-apocalypse-poison-put-hold/.
74. West, Cooper, and Armstrong, *Managing Wild Pigs*, 38–42.
75. "Hogzilla or Wilbur? Solving the Mystery Behind the Latest 'Giant Boar,'" *Field and Stream*, February 26, 2007, https://www.fieldandstream.com/photos/gallery/kentucky/2007/02/hogzilla-or-wilbur-solving-mystery-behind-latest-giant-boar.

## CHAPTER 2

1. Tom Moore Craig, ed., *Upcountry South Carolina Goes to War: Letters of the Anderson, Moore, and Brockman Families, 1853–1865* (Columbia: University of South Carolina Press, 2011), 52; Mecklin, cited in Winston Groom, *Shiloh,*

*1862* (Washington, DC: National Geographic, 2012), 339; Henry R. Pyne, *The History of the First New Jersey Cavalry* (Trenton, NJ: J.A. Beecher, 1871), 122; Peter Cozzens, *This Terrible Sound: The Battle of Chickamauga* (Urbana: University of Illinois Press, 1992), 517; Charles W. Porter, ed., *In the Devil's Dominions: A Union Soldier's Adventures in "Bushwhacker Country"* (Nevada, MO: Vernon County Historical Society, 1998), 42.

2. Drew Gilpin Faust, *This Republic of Suffering: Death and the American Civil War* (New York: Vintage, 2008), 59.

3. Gaiven Lowrie (1684), quoted in Robert R. Gradie III, "New England Indians and Colonizing Pigs," *Archives of the Papers of the Algonquian Conference* 15 (1984): 151.

4. Lewis Cecil Gray, *History of Agriculture in the Southern United States to 1860*, 2 vols. (Washington, DC: Carnegie Institute of Washington, 1933; repr., Gloucester, MA: Peter Smith, 1958), 1:38.

5. Peter O. Wacker and Paul G. E. Clemens, *Land Use in Early New Jersey: A Historical Geography* (Newark: New Jersey Historical Society, 1995), 62.

6. Joan Thirsk, *Food in Early Modern England: Phases, Fads, Fashions 1500–1760* (London: Hambledon Continuum, 2006), 242.

7. Gradie, "New England Indians," 152; Robert Malcolmson and Stephanos Mastoris, *The English Pig: A History* (Rio Grande, OH: Hambledon, 1998), 35–36, 35.

8. Darrett B. Rutman and Anita H. Rutman, *A Place in Time: Middlesex County, Virginia, 1650–1750* (New York: Norton, 1984), 45.

9. S. Max Edelson, *Plantation Enterprise in Colonial South Carolina* (Cambridge, MA: Harvard University Press, 2006), 41.

10. *American Husbandry*, vol. 1 (London: J. Bew, 1775), 23–24.

11. Lois Green Carr, Russell R. Menard, and Lorena S. Walsh, *Robert Cole's World: Agriculture and Society in Early Maryland* (Chapel Hill: University of North Carolina Press, 1991), 47.

12. William Byrd, *The Westover Manuscripts: Containing the History of the Dividing Line betwixt Virginia and North Carolina . . .* (Petersburg, VA: Edmund and Julian C. Ruffin, 1841), 16.

13. Carr, Menard, and Walsh, *Robert Cole's World*, 228–233.

14. Gray, *History of Agriculture*, 1:200.

15. Charles E. Brooks, *Frontier Settlement and Market Revolution: The Holland Land Purchase* (Ithaca, NY: Cornell University Press, 1996), 69.

16. John Woods, *Two Years' Residence in the Settlement on the English Prairie in the Illinois Country, United States* (London: Longman, Hurst, Rees, Orme, and Brown, 1822), 296.
17. Barbara Lawrence and Nedra Branz, eds., *The Flagg Correspondence, Selected Letters, 1816–1854* (Carbondale: Southern Illinois University Press, 1986), 18.
18. Malcolm Rohrbough, *The Trans-Appalachian Frontier: People, Societies, and Institutions, 1775–1850* (New York: Oxford University Press, 1978), 168–169; Matthew Foster letter, July 18, 1823, cited in "Early Nineteenth Century Agriculture" (unpublished manuscript, Conner Prairie, Fishers, IN).
19. Solon Robinson, "Notes of Travel in the Southwest, No. V," *Cultivator* 11(8) (August 1845): 239.
20. Mary Austin Holley, cited in Abigail Curlee, "A Texas Slave Plantation, 1831–1863," in *Plantation, Town, and County: Essays on the Local History of American Slave Society*, ed. Elinor Miller and Eugene D. Genovese (Urbana: University of Illinois Press, 1974), 315.
21. Cited in Forrest McDonald and Grady McWhiney, "From Self-Sufficiency to Peonage: An Interpretation," *American Historical Review* 85(5) (December 1980): 1106.
22. Frederick Law Olmsted, *A Journey through Texas: Or a Saddle-Trip on the Southwestern Frontier* (1857; Lincoln: University of Nebraska Press, 2004), 195.
23. Richard Parkinson, *A Tour in America in 1798, 1799, and 1800*, vol. 1 (London: J. Harding, 1805), 290.
24. Patrick Shirreff, *A Tour through North America* (Edinburgh: Oliver and Boyd, 1835), 249.
25. William Law Smith, quoted in Jack Temple Kirby, *Poquosin: A Study of Rural Landscape and Society* (Chapel Hill: University of North Carolina Press, 1995), 31.
26. James Stuart, *Three Years in North America*, 3rd ed. (Edinburgh: Robert Cadell, 1833), 91.
27. Cornelius Van Tienhoven, "Information Relative to Taking Up Lands in New Netherland" (1650), in *Remarkable Provinces: Readings in Early American History*, rev. ed., ed. John Demos (1972; Boston: Northeastern University Press, 1991), 25.
28. Towne and Wentworth exaggerated when they claimed that grain finishing with corn was common by 1700. Charles Wayland Towne and Edward Norris Wentworth, *Pigs: From Cave to Corn Belt* (Norman: University of Oklahoma Press, 1950), 86; Gray, *History of Agriculture*, 1:200–201; Virginia DeJohn Anderson,

*Creatures of Empire: How Domestic Animals Transformed Early America* (New York: Oxford University Press, 2004), 114.

29. David O. Percy, *Of Fast Horses, Black, Cattle, Woods Hogs, and Rat-Tailed Sheep: Animal Husbandry along the Colonial Potomac* (National Colonial Farm Research Report no. 4, 1979), 47; Lorena S. Walsh, *From Calabar to Carter's Grove: The History of a Virginia Slave Community* (Charlottesville: University Press of Virginia, 1997), 127; Mary V. Thompson, "Hogs," in *The Digital Encyclopedia of George Washington*, ed. Adam Shprintzen (Mount Vernon Estate, 2012), http://www.mountvernon.org/educational-resources/encyclopedia/hogs.

30. Esther Louise Larsen, "Pehr Kalm's Observations on the Fences of North America," *Agricultural History* 21(2) (April 1947): 75; James T. Lemon, *The Best Poor Man's Country: A Geographical Study of Early Southeastern Pennsylvania* (New York: Norton, 1972), 152–153, 157–158, 162; John G. Gagliardo, "German Agriculture in Pennsylvania," *Pennsylvania Magazine of History and Biography* 38(2) (April 1959): 197–198.; Amos Long, Jr., "Pigpens and Piglore in Rural Pennsylvania," *Pennsylvania Folklife* 19(2) (1960): 21.

31. Wacker and Clemens, *Land Use in Early New Jersey*, 187–188.

32. Wacker and Clemens, *Land Use in Early New Jersey*, 188.

33. Rebecca Burlend, *A True Picture of Immigration* (Lincoln: University of Nebraska Press, 1987),130.

34. William Cooper Howells, *Recollections of Life in Ohio, 1813–1840* (Cincinnati, OH: Robert Clarke, 1895), 65.

35. John Aston, "A History of Hogs and Pork Production in Missouri," *Monthly Bulletin of the Missouri State Board of Agriculture* 21(1) (January 1923): 41–42.

36. Dave Miles, "Paine Howard Diary Report, 1860–1864" (unpublished manuscript, Living History Farms, Urbandale, IA, n.d.).

37. Frederick Law Olmsted, *The Cotton Kingdom* (New York: Modern Library, 1984), 55.

38. C. M. Clay, "All about Hogs," *Illinois Farmer*, March 1859, 230.

39. Dana O. Jensen, ed., "I at Home: The Diary of a Yankee Farmer in Missouri, 1815–1816," *Missouri Historical Society Bulletin*, April 1957, 299; Miles, "Paine Howard Diary Report."

40. Jensen, "I at Home," 316.

41. Donald L. Winters, *Tennessee Farming, Tennessee Farmers: Antebellum Agriculture in the Upper South* (Knoxville: University of Tennessee Press, 1994), 122.

42. Henry W. Ellsworth, *The American Swine Breeder: A Practical Treatise on the Selection, Rearing, and Fattening of Swine* (Boston: Weeks, Jordan, 1840), 255–257.

43. "Swine," *Illinois Farmer*, June 1860, 87.
44. Lewis F. Allen, "Pork Feeding in the West," *Illinois Farmer*, July, 1857, 153.
45. Quotes from Ellsworth, *American Swine Breeder*, 251, 86–87, 247–249; Rensselaer County quote, "Hogs Fatted on Sweet Apples," *Hudson (OH) Observer and Telegraph*, October 25, 1832; Stuart, *Three Years in North America*, 420–421. Also see "Apples for Animals," *Genesee Farmer and Gardener's Journal*, April 27, 1839.
46. "Petition of Owners of Lands in the 'Neck' to Restrain Swine from Running at Large, 1703," *Pennsylvania Magazine of History and Biography* 28(2) (1904): 236–237.
47. James C. Bonner, *A History of Georgia Agriculture, 1732–1860* (Athens: University of Georgia Press, 1964), 30.
48. Frank E. Grizzard and D. Boyd Smith, *Jamestown Colony: A Political, Social, and Cultural History* (Santa Barbara, CA: ABC-CLIO, 2007), 90; Gray, *History of Agriculture*, 34; William Cronon, *Changes in the Land: Indians, Colonists, and the Ecology of New England* (New York: Hill & Wang, 1983), 136; Carl Bridenbaugh, *Fat Mutton and Liberty of Conscience: Society in Rhode Island, 1636–1690* (Providence, RI: Brown University Press, 1974), 40; Wacker and Clemens, *Land Use in Early New Jersey*, 66; Stewart, *"What Nature Suffers to Groe,"* 95; Wacker and Clemens, *Land Use in Early New Jersey*, 66.
49. Darett B. Rutman, *Husbandmen of Plymouth: Farms and Villages in the Old Colony, 1620–1692* (Boston: Beacon, 1967), 49; "Legislation Relating to Fences in Colonial Virginia" (Colonial Williamsburg Official History and Citizenship, 2018), http://history.org/history/teaching/fenceleg.cfm.
50. Cronon, *Changes in the Land*, 135: Wacker and Clemens, *Land Use in Early New Jersey*, 63: Carl J. Ekberg, *French Roots in the Illinois Country: The Mississippi Frontier in Colonial Times* (Urbana: University of Illinois Press, 1998), 136.
51. Timothy Silver, *A New Face on the Countryside: Indians, Colonists, and Slaves in South Atlantic Forests, 1500–1800* (Cambridge: Cambridge University Press, 1990), 179–180; Wacker and Clemens, *Land Use in Early New Jersey*, 66.
52. Brian Donahue, *The Great Meadow: Farmers and the Land in Colonial Concord* (New Haven, CT: Yale University Press, 2004), 96, 118, 174.
53. Pehr Kalm described triangular yokes in 1748. Larsen, "Pehr Kalm's Observations," 76; "An Act for the Better Regulating Swine," in *Acts and Laws Passed by the Great and General Court or Assembly of His Majesties Province of the Massachusetts-Bay in New England, May, 1731* (Boston: B. Green, 1732), 458.
54. Peter W. Cook, "Domestic Livestock of Massachusetts Bay, 1625–1725," in *The Dublin Seminar for New England Folklife: Annual Proceedings, 1986*, ed. Peter

Benes (Boston: Boston University, 1988), 121; Stephen Innes, *Labor in a New Land: Economy and Society in Seventeenth-Century Springfield* (Princeton, NJ: Princeton University Press, 1983), 87.

55. "An Act to Prevent Damages by Swine in the County of Westchester, Queens-County, and the County of Richmond," in *Acts Passed by the General Assembly of the Colony of New York* (New York: William Bradford, 1726), 67.
56. Gray, *History of Agriculture*, 2:835; Bonner, *History of Georgia Agriculture*, 145; Lynn A. Nelson, *Pharsalia: An Environmental Biography of a Southern Plantation, 1780–1880* (Athens: University of Georgia Press, 2007), 41.
57. Robert Leslie Jones, *History of Agriculture in Ohio to 1880* (Kent: Kent State University Press, 1983), 121.
58. Johann David Schoepf, *Travels in the Confederation, 1783–1784*, trans. and ed. Alfred J. Morrison (Philadelphia: William J. Campbell, 1911), 236.
59. R. Douglas Hurt, *The Ohio Frontier: Crucible of the Old Northwest, 1720–1830* (Bloomington: Indiana University Press, 1996), 212.
60. Claudia L. Bushman, *In Old Virginia: Slavery, Farming, and Society in the Journal of John Walker* (Baltimore: Johns Hopkins University Press, 2002), 129–132.
61. Ulrich Bonnell Phillips, *American Negro Slavery* (New York: Appleton, 1929), 59, 230.
62. Nelson, *Pharsalia*, 94; Steven Stoll, *Larding the Lean Earth: Soil and Society in Nineteenth-Century America* (New York: Hill & Wang, 2002), 125.
63. David J. Grettler, "Environmental Change and Conflict over Hogs in Early Nineteenth-Century Delaware," *Journal of the Early Republic* 19(2) (Summer 1999): 199, 204, 210, 216–217, 219.
64. John Taylor, *Arator: Being a Series of Agricultural Essays, Practical and Political*, Hogs, no. 45, 3rd ed. (Baltimore: John M. Carter, 1817), 134–135.
65. "The Loss Sustained by Farmers in Suffering Their Swine to Run at Large," *Medical and Agricultural Register*, August 1, 1806, 117.
66. Cited in Ellsworth, *American Swine Breeder*, 88.
67. "Swine Running at Large," *American Agriculturalist*, March 1845, 76.
68. Clarence Danhof, "The Fencing Problem in the Eighteen Fifties," *Agricultural History* 18(4) (October 1944): 176–177; Drew Addison Swanson, "Fighting over Fencing: Agricultural Reform and Antebellum Efforts to Close the Virginia Open Range," *Virginia Magazine of History and Biography* 117(2) (2009): 103–139; Gray, *History of Agriculture*, 2:843.
69. Sam Bowers Hilliard, *Hog Meat and Hoecake: Food Supply in the Old South* (Carbondale: Southern Illinois University Press, 1972), 93.

70. John Linton, "On Feeding Swine," *Saturday Magazine*, July 7, 1821, 18.
71. "On Rearing and Fattening Swine," *New England Farmer*, September 13, 49.
72. Ellsworth, *American Swine Breeder*, 104, 109, chap. 4.
73. Ellsworth, *American Swine Breeder*, 123.
74. "Management of Hogs," *Cultivator*, January 1840, 12; "Management of Hogs," *New England Farmer*, December 8, 1849, 26; "Management of Swine," *Spirit of the Times*, July 28, 1849, 23; "Management of a Stock of Hogs," *American Farmer*, October 1855, 117; "The Hog: Management in Iowa," cited in Earle D. Ross, *Iowa Agriculture* (Iowa City: State Historical Society of Iowa, 1951), 46.
75. "Hogs," *Illinois Farmer*, December 1858, 188.
76. Reynolds, "ART 166—Remarks upon the Treatment of Swine," *New-York Farmer and Horticultural Repository*, November 1, 1828, 269.
77. "Swine," *New England Farmer*, September 17, 1830, 68.
78. Francis S. Wiggins, *The American Farmer's Instructor* (Philadelphia: Orrin Rogers, 1840), 398.
79. "Pork with Little or No Corn," *Medical and Agricultural Register*, December 1, 1807, 372–373.
80. Ellsworth, *American Swine Breeder*, quotes on 224–225, discussion of feeds on 224–236.
81. "Thoughts on Raising and Feeding Swine," *American Museum, or Universal Magazine*, September 1, 1791, 123–124.
82. "On the Feeding of Hogs to Advantage," *American Museum, or Universal Magazine*, September 1, 1806, 139.
83. Ellsworth, *American Swine Breeder*, 155.
84. Cited in Ellsworth, *American Swine Breeder*, 157.
85. Wiggins, *The American Farmer's Instructor*, 398.
86. "Cooking Food for Hog," *Illinois Farmer*, January 1857, 8–9.
87. Ellsworth, *American Swine Breeder*, 68.
88. "Penning Hogs-Corn and Cob Meal-Manure-Growing Turnips," *American Agriculturalist*, April 1845, 125.
89. "Management of Swine at the South," *American Railroad Journal*, November 17, 1832, 743–744; "Hog Feeding and the Sweet Potato Culture in N.C.," *American Farmer*, November 1857, 166; Grady McWhiney, Warren O. Moore, Jr., and Robert F. Pace, eds., *"Fear God and Walk Humbly": The Agricultural Journal of James Mallory* (Tuscaloosa: University of Alabama Press, 1997), 56, 172.
90. Ellsworth, *American Swine Breeder*, 110–112.
91. "Swine," *New England Farmer*, September 17, 1830, 68.

92. *Genesee Farmer* quote in Ellsworth, *American Swine Breeder*, 84; Pennsylvania essay in Ellsworth, *American Swine Breeder*, 98–99.
93. Julian Wiseman, *A History of the British Pig* (London: Duckworth, 1986), chap. 3.
94. Juliet Clutton-Brock, "George Garrard's Livestock Models," *Agricultural History Review* 24(1) (1976), 18–21.
95. "Old Seed Corner," *Bloomfield Democratic Clarion*, December 7, 1859, cited in Mildred Throne, "'Book Farming' in Iowa, 1840–1870," *Iowa Journal of History* 49(2) (April 1951): 117–142.

## CHAPTER 3

1. *The Diary of Elisabeth Koren, 1853–1855*, trans. and ed. David T. Nelson (Northfield, MN: Norwegian-American Historical Association, 1955), 234, 100, 156, 169.
2. Glenda Riley, ed., "A Prairie Diary: Mary St. John," in *Prairie Voices: Iowa's Pioneering Women* (Ames: Iowa State University Press, 1996), 68–69.
3. Joan Thirsk, *Food in Early Modern England: Phases, Fads, Fashions 1500–1760* (London: Hambledon Continuum, 2006), 169–170.
4. Christopher Dyer, "Changes in Diet in the Late Middle Ages: The Case of Harvest Workers," in *Everyday Life in Medieval England*, ed. Christopher Dyer (1994; London: Hambledon and London, 2000), 86, 89, 93; Julian Wiseman, *A History of the British Pig* (London: Duckworth, 1986), chaps. 1 and 2; U. Albarella, "Pig Husbandry and Pork Consumption in Medieval England," in *Food in Medieval England: Diet and Nutrition*, ed. C. M. Woolgar, D. Serjeantson, and T. Waldren (Oxford: Oxford University Press, 2006); Christopher Dyer, "English Diet in the Later Middle Ages," in *Social Relations and Ideas: Essays in Honour of R.H. Hilton*, ed. T. H. Aston, P. R. Cross, Christopher Dyer, and Joan Thirsk (Cambridge: Cambridge University Press, 1984), 206; Brett Mizelle, *Pig* (London: Reaktion Books, 2011), chap. 1; C. M. Woolgar, "Meat and Dairy Products in Late Medieval England," in Woolgar, Serjeantson, and Waldren, *Food in Medieval England*.
5. Adam Fox, "Food, Drink and Social Distinction in Early Modern England," in *Remaking English Society: Social Relations and Social Change in Early Modern England*, ed. Steven Hindle, Alexandra Shepard, and John Walter (Woodbridge, UK: Boydell Press, 2013), 166; David Hackett Fisher, *Albion's Seed: Four British Folkways in America* (Oxford: Oxford University Press, 1989), 729; Thirsk, *Food in Early Modern England*, 204–205, 241, 249.

6. James G. Parsons, "The Acorn-Hog Economy of the Oak Woodlands of Southwestern Spain," *Geographical Review* 52(2) (April 1962): 211–235; Rebecca Earle, *The Body of the Conquistador: Food, Race, and the Colonial Experience in Spanish North America, 1492–1700* (Cambridge: Cambridge University Press, 2012), 60–61, 67, 73–75; William W. Dunmire, *New Mexico's Spanish Livestock Heritage: Four Centuries of Animals, Land, and People* (Albuquerque: University of New Mexico Press, 2013), 22.

7. Marcel Amills, Oscar Ramirez, Ofelia Galman-Omitogun, and Alex Clop, "Domestic Pigs in Africa," *African Archaeological Review* 30 (2013): 73–82; Frederick Douglass Opie, *Hog and Hominy: Soul Food from Africa to America* (New York: Columbia University Press, 2008), 4; Lisa R. Shiflett, "West African Food Traditions in Virginia Foodways: A Historical Analysis of Origins and Survivals" (MA thesis, East Tennessee State University, 2004), 37, https://dc.etsu.edu/etd/925/.

8. John Shelton Reed, "There's a Word for it—The Origins of 'Barbecue,'" *Southern Cultures* 13(4) (Winter 2007): 138–146.

9. Rhys Isaac, *The Transformation of Virginia, 1740–1790* (New York: Norton, 1988), 79, 85.

10. Sandra L. Oliver, *Food in Colonial and Federal America* (Westport, CT: Greenwood, 2005), 186.

11. Henry M. Miller, "An Archaeological Perspective on the Evolution of Diet in the Colonial Chesapeake, 1620–1745," in *Colonial Chesapeake Society*, ed. Lois Green Carr, Philip D. Morgan, and Jean B. Russo (Chapel Hill: University of North Carolina Press, 1988), 181–186.

12. See Figure 1, "Livestock Holdings in St. Mary's County, Maryland, 1638–1705," in Miller, "Archaeological Perspective," 179.

13. Miller, "Archaeological Perspective," 185, 190; Paul G. E. Clemens, "The Operation of an Eighteenth-Century Tobacco Plantation," *Agricultural History* 49(3) (July 1975): 525.

14. William Byrd, *The Prose Works of William Byrd of Westover: Narratives of a Colonial Virginian*, ed. Louis B. Wright (Cambridge, MA: Belknap, 1966), 185; William Byrd, "The Secret History of the Line by William Byrd II," in *Colonial American Travel Narratives*, ed. Wendy Martin (New York: Penguin, 1994), 96–97.

15. Lorena S. Walsh, "Feeding the Eighteenth-Century Town Folk, or Whence the Beef," *Agricultural History* 73(3) (Summer 1999): 269; Miller, "Archaeological Perspective," 191.

16. Andrew Burnaby, *Travels through the Middle Settlements in North America in the Years 1759 and 1760* (London: T. Payne, 1798), 16.
17. "A Letter from Mr. John Clayton, Rector of Crofton at Wakefield in Yorkshire, to the Royal Society," May 12, 1688 (Virtual Jamestown Project), http://etext.lib.virginia.edu/etcbin/jamestown-browse?id=J1074.
18. Burnaby, *Travels through the Middle Settlements*, 63.
19. James Stuart, *Three Years in North America*, 3rd ed. (Edinburgh: Robert Cadell, 1833), 104.
20. Quoted in Sarah F. McMahon, "A Comfortable Subsistence: Changes in the New England Diet, 1620–1840," *William and Mary Quarterly* 42(1) (January 1985): 34.
21. McMahon, "Comfortable Subsistence," 55–56.
22. Michael D. and Sophie D. Coe, "Mid-Eighteenth-Century Food and Drink on the Massachusetts Frontier," in *Foodways in the Northeast: The Dublin Seminar for New England Folklife: Annual Proceedings 1983*, ed. Peter Benes (Boston: Boston University, 1984), 42; Daphne L. Derven, "Wholesome, Toothsome, and Diverse: Eighteenth-Century Foodways in Deerfield, Massachusetts," in Benes, *Foodways in the Northeast*, 55–57; Steven R. Pendery, "The Archeology of Urban Foodways in Portsmouth, New Hampshire," in Benes, *Foodways in the Northeast*; Joanne Bowen, "A Comparative Analysis of the New England and Chesapeake Herding Systems," in *Historical Archaeology of the Chesapeake*, ed. Paul A. Shackel and Barbara J. Little (Washington, DC: Smithsonian Institution Press, 1994), 157.
23. David B. Landon, "Feeding Colonial Boston: A Zooarchaeological Study," *Historical Archaeology* 30(1) (1996): 44–45.
24. James T. Lemon, *The Best Poor Man's Country: A Geographical Study of Early Southeastern Pennsylvania* (New York: Norton, 1972), 61, 62, 68; Michael V. Kennedy, "'Cash for His Turnips': Agricultural Production for Local Markets in Colonial Pennsylvania, 1725–1783," *Agricultural History* 74(3) (Summer 2000): 600–601; Richard J. Hooker, *Food and Drink in America: A History* (Indianapolis, IN: Bobbs-Merrill, 1981), 41; Sarah Sportman, Craig Cipolla, and David Landon, "Zooarchaeological Evidence for Animal Husbandry and Foodways at Sylvester Manor," *Northeast Historical Archaeology* 36(1) (2007): 136–137; Haskel J. Greenfield, "From Pork to Mutton: A Zooarchaeological Perspective on Colonial New Amsterdam and Early New York City," *Northeast Historical Archaeology* 18 (1989): 85–110.
25. Michael Olmert, "Smokehouses," *Colonial Williamsburg Journal*, Winter

2004–2005, http://www.history.org/foundation/journal/winter04-05/smoke.cfm.

26. Carl Bridenbaugh, ed., *Gentleman's Progress: The Itinerarium of Dr. Alexander Hamilton, 1744* (Chapel Hill: University of North Carolina Press, 1948), 169.

27. "The Journal of Madam Knight," in Martin, *Colonial American Travel Narratives*, 54.

28. Bridenbaugh, *Gentleman's Progress*, 87, 94, 96.

29. James McWilliams, *A Revolution in Eating: How the Quest for Food Shaped America* (New York: Columbia University Press, 2005), 167.

30. McWilliams, *Revolution in Eating*, 118.

31. Lorena S. Walsh, "Consumer Behavior, Diet, and the Standard of Living in Late Colonial and Early Antebellum America, 1770–1840," in *American Economic Growth and Standards of Living before the Civil War*, ed. Robert E. Gallman and John J. Wallis (Chicago: University of Chicago Press, 1993), 218; *Harper's Weekly* quote from Susan Williams, *Savory Suppers and Fashionable Feasts: Dining in Victorian America* (Knoxville: University of Tennessee Press, 1996), 100–101.

32. The importance of pork on the frontier is addressed in Reginald Horsman, *Feast or Famine: Food and Drink in American Westward Expansion* (Columbia: University of Missouri Press, 2008). Michael D. Thompson suggested that pork consumption increased in the US Southeast in the early nineteenth century in "'Everything but the Squeal': Pork as Culture in Eastern North Carolina," *North Carolina Historical Review* 82 (October 2005): 466; Thomas A. Woods, "A Portal to the Past: Property Taxes in the Kingdom of Hawai'i," *Hawaiian Journal of History* 45 (2011): 1–47.

33. Benjamin Henry Latrobe, *The Journal of Latrobe*, repr. ed. (Bedford, MA: Applewood Books, 1876), 15–17.

34. Carpenter quote in Jacqueline Williams, "Food on the Oregon Trail," *Overland Journal* 11(2) (Summer 1993): 4; Albert Edward Belanger, "'Equipping Ourselves in Every Possible Way': Contents of Wagons on the 1851 Oregon Trail," *Overland Journal* 24(4) (Winter 2006): 144; Randolph B. Marcy, *The Prairie Traveler: A Hand Book for Overland Expeditions* (1859; repr., Cambridge, MA: Applewood Books, 1988), 36.

35. Eugene M. Hattori and Jerre L. Kosa, "Packed Pork and Other Foodstuffs from the California Gold Rush," in *The Hoff Store Site and Gold Rush Merchandise from San Francisco, California*, Special Publication Series No. 7, ed. Allen G. Pastron and Eugene M. Hattori (Ann Arbor, MI: Society for Historical

Archaeology, 1990), 83–87; Chauncy L. Canfield, ed., *The Diary of a Forty-Niner* (Boston: Houghton Mifflin, 1920), 3.

36. Peter Kolchin, *American Slavery, 1619–1877* (New York: Hill & Wang, 1993), 113; James O. Breeden, *Advice among Masters: The Ideal in Slave Management in the Old South* (Westport, CT: Greenwood, 1980), 89, 200; James W.C. Pennington, *The Fugitive Blacksmith; or, Events in the History of James W.C. Pennington*, 2nd ed. (London: Charles Gilpin, 1849), 65; Richard Russell, *North America, Its Agriculture and Climate* (Edinburgh: Adam and Charles Black, 1857), 266; Frederick Law Olmsted, *The Cotton Kingdom* (New York: Modern Library, 1984), 433; Sam Bowers Hilliard, *Hog Meat and Hoe Cake: Food Supply in the Old South, 1840–1860* (Carbondale: Southern Illinois University Press, 1972), 104. Wilma Dunaway argued that slaves in the Appalachian region "probably had a much less diverse meat diet than slaves in other parts of the South." See *The African-American Family in Slavery and Emancipation* (Cambridge: Cambridge University Press, 2003), 103.

37. Joe Gray Taylor, *Eating, Drinking, and Visiting in the South: An Informal History* (Baton Rouge: Louisiana State University Press, 1982), 26.

38. Walsh, "Consumer Behavior," 245; Lu Ann De Cunzo, *A Historical Archaeology of Delaware: People, Contexts, and the Cultures of Agriculture* (Knoxville: University of Tennessee Press, 2004), 65.

39. Latrobe, *Journal of Latrobe*, 34.

40. Olmsted, *Cotton Kingdom*, 71, 139, 231, 305, 325.

41. John Hamilton Cornish diary, cited in Hooker, *Food and Drink in America*, 171.

42. John Woods, *Two Years' Residence in the Settlement on the English Prairie in the Illinois Country, United States* (London: Longman, Hurst, Rees, Orme, and Brown, 1822), 37; Stuart, *Three Years in North America*, 158; Martineau quoted in Taylor, *Eating, Drinking, and Visiting*, 57.

43. Cited in Thompson, "'Everything but the Squeal,'" 467.

44. Johann David Schoepf, *Travels in the Confederation, 1783–1784*, trans. and ed. Alfred J. Morrison (Philadelphia: William J. Campbell, 1911), 166, 236.

45. Reuben Gold Thwaites, ed., *Travels West of the Alleghanies Made in 1793–96 by André Michaux; in 1802 by F.A. Michaux, and in 1803 by Thaddeus Mason Harris, M.A.* (Cleveland: Arthur H. Clark, 1904), 139, 147, 190.

46. Emma S. Vonnegut, trans. and ed., *The Schramm Letters: Written by Jacob Schramm and Members of His Family from Indiana to Germany in the Year 1836* (1935; repr., Indianapolis: Indiana Historical Society, 2005), 67–68.

47. William Cooper Howells, *Recollections of Life in Ohio, 1813–1840* (Cincinnati, OH: Robert Clarke, 1895), 140, 148.
48. Ralph H. Bowen, ed., *A Frontier Family in Minnesota: The Letters of Theodore and Sophie Bost, 1851–1920* (Minneapolis: University of Minnesota Press, 1981), 92.
49. "Turkey from a Hog: Omaha's First Christmases," *History Nebraska*, December 17, 2013, https://history.nebraska.gov/blog/turkey-hog-omaha%E2%80%99s-first-christmases.
50. William M. Selby, Michael J. O'Brien, and Lynn M. Snyder, "The Frontier Household," in *Grassland, Forest, and Historical Settlement: An Analysis of Dynamics in Northeast Missouri*, ed. Michael J. O'Brien (Lincoln: University of Nebraska Press, 1984), 307–313.
51. Asa Sheldon, *Yankee Drover: Being the Unpretending Life of Asa Sheldon, Farmer, Trader, and Working Man, 1788–1870* (Hanover, NH: University Press of New England, 1988), 25–26.
52. Hooker, *Food and Drink in America*, 113, 169.
53. Williams, *Savory Suppers*, 100–101.
54. Arnold Schrier, "A Russian Observer's Visit to 'Porkopolis,' 1857," *Cincinnati Historical Society Bulletin* 29 (Spring 1971): 33–35.
55. Leslie cited in Hooker, *Food and Drink in America*, 112–113.
56. "Swine's Flesh," *Illinois Farmer*, May 1859, 266–267.
57. Bill I. Wiley, *The Life of Billy Yank: The Common Soldier of the Union* (Baton Rouge: Louisiana State University Press, 1952), 239; Warren Hapgood Freeman and Eugene Harrison Freeman, *Letters from Two Brothers Serving in the War for the Union to Their Family at Home* (Cambridge, MA, 1871), 19.
58. Cited in Joe A. Mobley, *Weary of War: Life on the Confederate Home Front* (Westport, CT: Greenwood, 2008), 22.
59. James Mallory, *"Fear God and Walk Humbly": The Agricultural Journal of James Mallory*, ed. Grady McWhiney, Warren O. Moore, Jr., and Robert F. Pace (Tuscaloosa: University of Alabama Press, 1997), 326–327.
60. Paul W. Gates, *Agriculture and the Civil War* (New York: Knopf, 1965), 90–91.
61. John Solomon Otto, *Southern Agriculture during the Civil War Era, 1860–1880* (Westport, CT: Greenwood, 1994), 38.
62. Tom Moore Craig, ed., *Upcountry South Carolina Goes to War: Letters of the Anderson, Brockman, and Moore Families, 1853–1865* (Charleston: University of South Carolina Press, 2011), 100.

63. Cited in Taylor, *Eating, Drinking, and Visiting*, 95.
64. Taylor, *Eating, Drinking, and Visiting*, 95.
65. Ella Lonn, *Salt as a Factor in the Confederacy*, Southern Historical Publications No. 4 (Tuscaloosa: University of Alabama Press, 1965), 36–41, 36.
66. William John Grayson and Elmer L. Puryear, "The Confederate Diary of William John Grayson, Continued," *South Carolina Historical Magazine* 63(4) (October 1962): 224.
67. James Dugan, *History of Hurlbut's Fighting Fourth Division: And Especially the Marches, Toils, Privations, Adventures, Skirmishes, and Battles of the Fourteenth Illinois Infantry* (Cincinnati, OH: E. Morgan, 1863), 205.
68. John G. B. Adams, *Reminiscences of the Nineteenth Massachusetts Regiment* (Boston: Write and Potter, 1899), 136.
69. Cited in Lisa M. Brady, "Wilderness of War: Nature and Strategy in the American Civil War," *Environmental History* 10 (Summer 2005): 436.
70. Kenneth Lyftogt, *Left for Dixie: The Civil War Diary of John Rath* (Parkersburg, IA: Mid-Prairie Books, 1991), 61, 64.
71. Joseph T. Glatthaar, *The March to the Sea and Beyond: Sherman's Troops in the Savannah and Carolinas Campaigns* (New York: New York University Press, 1985), 127.
72. *OR*, ser. 1, vol. 39, pt. 2, 830, cited in Dunaway, *African-American Family*, 181.
73. *OR*, ser. 2, vol. 5, 601.
74. Craig, *Upcountry South Carolina Goes to War*, 90.
75. Glatthaar, *March to the Sea*, 125.
76. Dorothy G. Mahanes and Wallace S. Mahanes, "National Register of Historic Places Nomination/Inventory: Northbank" (2005), 7, http://www.dhr.virginia.gov/registers/Counties/KingandQueen/049-0051_Northbank_2006_NR_final.pdf.
77. Craig, *Upcountry South Carolina Goes to War*, 82.
78. Letter from James H. Meteer to Caleb Mills, September 13, 1864, Indiana Historical Society, http://images.indianahistory.org/cdm4/document.php?CISOROOT=/dc008&CISOPTR=391&REC=5.
79. Leo Huff, ed., *The Civil War Letters of Albert Demuth and Roster, Eighth Missouri Volunteer Cavalry* (Springfield, MO: Greene County Historical Society, 1997), 55.

## CHAPTER 4

1. "To the Honorable the Corporation of the City of New-York, the Memorial of the Swine of Belonging to Said City Most Humbly Representeth," *American Magazine*, January 1788, 1, 2.
2. "To the Owners of Swine," *Tickler*, April 5, 1809, 8. For examples of sick and

aggressive hogs, see "A Distressing Accident . . . ," *Daily National Intelligencer*, May 9, 1829, column E; "To the Editor of the Herald," *Cleveland Herald*, June 7, 1848, column A; "Satan in a Swine," *Daily Evening Bulletin*, January 4, 1859, column B.

3. Joel A. Tarr, "The Metabolism of the Industrial City," *Journal of Urban History* 28 (July 2002): 511–545; "Thoughts on the Law Prohibiting Hogs to Prowl the Streets of Philadelphia," *American Museum*, August 1, 1788, 133.
4. "Thoughts on the Law Prohibiting Hogs to Prowl," 133.
5. Catherine McNeur, *Taming Manhattan: Environmental Battles in the Antebellum City* (Cambridge, MA: Harvard University Press, 2014); McNeur, "The 'Swinish Multitude': Controversies over Hogs in Antebellum New York City," *Journal of Urban History* 37(3) (2011): 642.
6. Charles Henry Wilson, *The Wanderer in America; or, Truth at Home* (Thirsk: Henry Masterman, 1823) 18; Ræder cited in Ted Steinberg, *Down to Earth: Nature's Role in American History* (Oxford: Oxford University Press, 2009), 158; Dickens cited in Patricia L. Dooley, *The Early Republic: Primary Documents on Events from 1799 to 1820* (Westport, CT: Greenwood, 2004), 305.
7. This case is cited in Howard B. Rock, "A Delicate Balance: The Mechanics and the City in the Age of Jefferson," *New York Historical Society Quarterly* 63 (April 1979): 101.
8. Stonington quote, *Providence Patriot, Columbian Phenix*, May 11, 1825, column E; New York quotes, "One Grand Piggery," *Universalist Watchman, Repository and Chronicle*, November 13, 1830, 116.
9. "An Ordinance Prohibiting Swine from Running at Large in the Town of Chillicothe," *Scioto Gazette*, April 24, 1802, column D; Editorial, *Gazette of the United States*, August 12, 1803, 210, column C; "An Act to Prevent Swine from Rooting or Otherwise Destroying the Pastures of the City of Washington," *National Intelligencer and Washington Advertiser*, no. 1324 (April 5, 1809): column A.
10. "A Law Respecting Swine," *National Advocate*, October 25, 1817, column E; Hendrik Hartog, "Pigs and Positivism," *Wisconsin Law Review* 4 (July 1985): 2–3.
11. Hartog, "Pigs and Positivism," 3–6, 6.
12. Edwin G. Burrows and Mike Wallace, *Gotham: A History of New York to 1898* (Oxford: Oxford University Press, 1999), 477; Hartog, "Pigs and Positivism," 10.
13. "The Cholera and The Hogs," *Brooklyn Daily Eagle*, May 28, 1849, 3.

14. "Pig War," *Brooklyn Daily Eagle*, August 6, 1849, 3; "The Little Pig's Petition to His Honor the Mayor," *Brooklyn Daily Eagle*, September 8, 1849, 3.
15. "Reforms and the Pound," *New York Times*, June 30, 1854; "The Pig Drivers' Troubles," *Brooklyn Daily Eagle*, August 6, 1856, 3.
16. "The Offal and Piggery Nuisances," *New York Times*, July 27, 1859; Burrows and Wallace, *Gotham*, 785–787; "The Public Health," *New York Herald*, July 14, 1859, 5, column C; "The Hog War—A Policeman in a Tight Place," *New York Herald*, August 11, 1859, 2, column F.
17. "Results of the War upon the Piggeries," *New York Times*, September 20, 1859; "Swine in New York," *Boston Daily Advertiser*, September 21, 1859.
18. McNeur, "'Swinish Multitude,'" 652, 640; Burrows and Wallace, *Gotham*, 747.
19. "Help Yourselves to Pork," *Scientific American* 2 (33) (May 8, 1847): 258; *Daily National Intelligencer*, August 5, 1850, 679, column D; *Maine Farmer*, October 28, 1852, cited in Clarence Danhof, "The Fencing Problem in the Eighteen Fifties," *Agricultural History* 18(4) (October 1944): 177; "How to Deal with Goats and Swine Running at Large," *Daily Evening Bulletin*, January 14, 1858, column E.
20. John Duffy, "Hogs, Dogs, and Dirt: Public Health in Early Pittsburg," *Pennsylvania Magazine of History and Biography* 87 (3) (July 1963): 299–301.
21. John B. Blake, *Public Health in the Town of Boston, 1630–1822* (Cambridge, MA: Harvard University Press, 1959), 167–169.
22. "Boston Piggery," *Genesee Farmer*, April 29, 1837, 136.
23. Marc Linder and Lawrence S. Zacharias, *Of Cabbages and Kings County: Agriculture and the Formation of Modern Brooklyn* (Iowa City: University of Iowa Press, 1999), 252–253.
24. Linder and Zacharias, *Of Cabbages and Kings County*, 255–256; "Raiding the Piggeries," *Brooklyn Daily Eagle*, June 4, 1885, 4.
25. "Pigs and Politics," *Boston Daily Globe*, November 2, 1893, 8.
26. "City Drainage—The Keeping of Swine," *Bangor Daily Whig and Courier*, July 10, 1866.
27. "The Hogs Must Go," *Milwaukee Daily Journal*, July 29, 1884.
28. Mathew R. Walpole, "The Closing of the Open Range in Watauga County, N.C.," *Appalachian Journal* 16 (4) (Summer 1989): 323.
29. Martin V. Melosi, *The Sanitary City: Environmental Services in the American City from Colonial Times to the Present*, abridged ed. (Pittsburgh: University of Pennsylvania Press, 2008), 163.

30. "Want Dumping Place in Long Beach," *Los Angeles Times*, December 25, 1901, 13; "Garbage Disposal," *Los Angeles Times*, March 17, 1903, A2.
31. "What's to Be Done with City's Garbage? Tense Question Demands Quick Answer," *Los Angeles Times*, November 6, 1910, I11; "Garbage Matter Gives Offense," *Los Angeles Times*, April 12, 1911, I14; "City Officials Sniff Garbage," *Los Angeles Times*, April 15, 1911, I12.
32. "Rotten Garbage Swine's Feast," *Los Angeles Times*, July 8, 1911, I17; "Public Inhales Typhoid Germs," *Los Angeles Times*, August 29, 1911, I14: "Reducing Plant Recommended," *Los Angeles Times*, November 2, 1912, II10.
33. US Food Administration, *Garbage Utilization with Particular Reference to Utilization by Feeding* (Washington, DC: Government Printing Office, 1918), 3–4.
34. "Garbage for Hogs," *Wallaces' Farmer*, September 28, 1917, 6.
35. E. A. Cahill, "Control of Hog Cholera with Particular Application to Garbage Feeding Plants," *Berkshire World and Corn Belt Stockman*, June 1, 1918, 56.
36. Cahill, "Control of Hog Cholera," 56; US Food Administration, *Garbage Utilization*, 18; Raymond R. Birch, *Hog Cholera: Its Nature and Control* (New York: Macmillan, 1922), 18.
37. "Council O.K.'s Garbage Deal," *Los Angeles Times*, July 31, 1921, I7.
38. "Fontana Hogs to Get Los Angeles Garbage," *San Bernardino County Sun*, August 2, 1921, 7; "Fontana Farms Co. Excavating for Hog Ranch to Be at Declez," *San Bernardino County Sun*, September 2, 1921, 13.
39. "Swine Project Record in Size," *Los Angeles Times*, April 3, 1927, E11.
40. Edward N. Wentworth and Tage U. H. Ellinger, *Progressive Hog Raising* (Chicago: Armour's Livestock Bureau, 1926), 79; quotes from F. G. Ashbrook and J. D. Bebout, "Disposal of City Garbage by Feeding to Hogs," US Department of Agriculture Circular No. 80 (Washington, DC: Government Printing Office, December 1917), 8.
41. Walter H. Stoling, *Food Waste Materials: A Survey of Urban Garbage Production, Collection, and Utilization* (Washington, DC: US Department of Agriculture, Bureau of Agricultural Economics, September 1941), 7–8.
42. Charles W. Snyder, "Effects of Sewage on Cattle and Garbage on Hogs," *Sewage and Industrial Wastes* 23 (10) (October 1951): 1239; Wallace Sullivan, Earl Maharg, and E. H. Hughes, "The Garbage Hog Feeding Business in California," Circular No. 166 (Berkeley: Agricultural Extension Service, College of Agriculture, University of California, April 1950), 4–5.
43. Snyder, "Effects of Sewage," 1239–1240.

44. E. R. Baumann and M. T. Skodje, "Garbage Cooking in Iowa," July 31, 1954, Iowa State University Library, Special Collections Department, R. A. Packer Collection, RS 14/6/11, box 7 folder 9.
45. Jay B. Shumaker, James E. Harbottle, and Myron G. Schultz, "Recent Outbreaks of Trichinosis in the United States," *Journal of Infectious Diseases* 120(3) (September 1969): 396; Andrew Moorhead, Paul E. Grunenwald, Vance J. Dietz, and Peter M. Schantz, "Trichinellosis in the United States, 1991–1996: Declining but Not Gone," *American Journal of Tropical Medicine and Hygiene* 60(1) (1999): 69.
46. "Waste Food Utilization," Report on National Conference on Maintaining Swine Health Regulations, March 20, 1972, Iowa State University Library, Special Collections Department, Neal Black Papers, MS-78, box 2, folder 12.
47. M. L. Westendorf and R. O. Myer, "Feeding Food Wastes to Swine," AS143 (Animal Sciences Department, Florida Cooperative Extension Service, Institute of Food and Agricultural Sciences, University of Florida, May 2004, Reviewed August 2012), 1–2.
48. Ray Zeman, "Garbage Disposal Confounds County," *Los Angeles Times*, December 3, 1950, 40.
49. Zeman, "Garbage Disposal Confounds County," 40.
50. "Piggeries Draw Fire at Hearing, *Hartford Courant*, July 31, 1942, 18.
51. "Complaints Are Caused by Garbage," *Hartford Courant*, June 19, 1945, 14; "Garbage Collections," *Hartford Courant*, August 28, 1947, 10.
52. "Martin Issues Ultimatums on Garbage Disposition," *Hartford Courant*, June 7, 1952, 6A; "Health Officer to Serve Order to Stop Nuisance," *Hartford Courant*, June 13, 1952, 22A; "50 Attend Public Meeting to 'Air' Garbage Situation," *Hartford Courant*, September 20, 1952, 6A.
53. "Health Director to Check Improvements at Piggery," *Hartford Courant*, June 21, 1958, 8A; "Martin Tightens Rules for Piggery Dumping," *Hartford Courant*, August 7, 1958; "Inspection Shows Area of Piggery Cleaned Up," *Hartford Courant*, July 12, 1960, 12A; "Order Closing Piggery to Affect Wethersfield," *Hartford Courant*, February 11, 1961, 9A.
54. "Four Pig Stys Moved on Health Dept. Order," *Hartford Courant*, April 1, 1954, 22; Roger Dove, "Pigs vs. People—Who's in Whose Yard?," *Hartford Courant*, April 4, 1954, A1.
55. Gerald J. Demeusy, "Court Banishes Pigs from East Hartford," *Hartford Courant*, June 24, 1954, 1.
56. "Controversial Pigs Root Again in East Hartford," *Hartford Courant*, March

15, 1956, 1; "Round Two Scheduled in 'Rights of Pigs' Case," *Hartford Courant*, February 6, 1958, 19.

57. Troy Paff, "Bob Combs: Pig Farmer," *Journeyman Project*, March 14, 2014, http://blog.journeymanproject.net/2014/03/bob-combs-pig-farmer/.

58. Steven Green, "NLV Homebuyers Sue Builder over Pig Farm Odors," *Las Vegas Sun*, September 4, 2009, http://lasvegassun.com/news/2009/sep/04/nlv-homebuyers-sue-builder-over-pig-farm-odors/; Judy Smith, "The Pig Farm Is Hardly a Secret," *Las Vegas Sun*, May 2, 2013, http://lasvegassun.com/news/2013/may/02/pig-farm-hardly-secret/.

59. Eli Segal, "North Las Vegas Farmer Who Earned His Chops in Pig Business Sells Out," *North Las Vegas Review-Journal*, November 16, 2016.

## CHAPTER 5

1. John J. McCusker and Russell R. Menard, *The Economy of British North America, 1607–1789* (Chapel Hill: University of North Carolina Press, 1985), 145, 170, 174, 199; Ross H. Cordy, "The Effects of European Contact on Hawaiian Agricultural Systems, 1778–1819," *Ethnohistory* 19 (4) (Fall 1972): 400–402.

2. Maverick quoted in Robert R. Gradie III, "New England Indians and Colonizing Pigs," *Archives of the Papers of the Algonquian Conference* 15 (1984): 154; John Josselyn, *An Account of Two Voyages to New-England, Made during the Years 1638, 1663* (Boston: William Veazie, 1865), 162; Carl Bridenbaugh, *Fat Mutton and Liberty of Conscience: Society in Rhode Island, 1636–1690* (Providence, RI: Brown University Press, 1974), 41; *American Husbandry*, vol. 1 (London: J. Bew, 1775), 56.

3. Bridenbaugh, *Fat Mutton*, 41; Carl Bridenbaugh, ed., *Gentleman's Progress: The Itinerarium of Dr. Alexander Hamilton, 1744* (Chapel Hill: University of North Carolina Press, 1948), 29; Rudolf A. Clemen, *The American Livestock and Meat Industry* (New York: Ronald Press, 1923), 26; Andrew Burnaby, *Travels through the Middle Settlements in North America in the Years 1759 and 1760* (London: T. Payne, 1798), 26.

4. Paul G. E. Clemens, "The Operation of an Eighteenth-Century Tobacco Plantation," *Agricultural History* 49(3) (July 1975): 527.

5. Howard S. Russell, *A Long, Deep Furrow: Three Centuries of Farming in New England* (Hanover, NH: University Press of New England, 1976), 83; Lorena S. Walsh, *From Calabar to Carter's Grove: The History of a Virginia Slave Community* (Charlottesville: University Press of Virginia, 1997), 120.

6. Stephen Innes, *Labor in a New Land: Economy and Society in Seventeenth-Century Springfield* (Princeton, NJ: Princeton University Press, 1983), 30–33.
7. Carl J. Ekberg, *French Roots in the Illinois Country: The Mississippi Frontier in Colonial Times* (Urbana: University of Illinois Press, 1998), 209–210.
8. Clemen, *American Livestock and Meat Industry*, 58–59.
9. Robert Leslie Jones, *History of Agriculture in Ohio to 1880* (Kent: Kent State University Press, 1983), 122–123.
10. Howard Johnson, *A Home in the Woods, Pioneer Life in Indiana: Oliver Johnson's Reminiscences of Early Marion County*, repr. ed. (Bloomington: Indiana University Press, 1991), 111.
11. Johnson, *Home in the Woods*, 112.
12. Johnson, *Home in the Woods*, 113–114.
13. Lewis Cecil Gray, *History of Agriculture in the Southern United States to 1860*, 2 vols. (Washington, DC: Carnegie Institute of Washington, 1933; repr., Gloucester, MA: Peter Smith, 1958), 2:840.
14. Donald L. Winters, *Tennessee Farming, Tennessee Farmers: Antebellum Agriculture in the Upper South* (Knoxville: University of Tennessee Press, 1994), 65; quote from Edmund Cody Burnett, "Hog Raising and Hog Driving in the Region of the French Broad River," *Agricultural History* 20(2) (April 1946): 88.
15. John Mack Faragher, *Sugar Creek: Life on the Illinois Prairie* (New Haven, CT: Yale University Press, 1986), 103; "To Market with Hogs," *Palimpsest* 12 (February 1931): 57–59.
16. Robert Russell, *North America, Its Agriculture and Climate* (Edinburgh: Adam and Charles Black, 1857), 89.
17. Wilson J. Warren, *Tied to the Great Packing Machine: The Midwest and Meatpacking* (Iowa City: University of Iowa Press, 2007), 40–46; Shane Hamilton, *Trucking Country: The Road to America's Wal-Mart Economy* (Princeton, NJ: Princeton University Press, 2008).
18. R. Douglas Hurt, "Pork and Porkopolis," *Cincinnati Historical Society Bulletin* 40 (Fall 1992): 196; John E. Stealey III, *The Antebellum Kanawha Salt Business and Western Markets* (Lexington: University Press of Kentucky, 1993), chap. 4.
19. Margaret Walsh, *The Rise of the Midwestern Meat Packing Industry* (Lexington: University Press of Kentucky, 1982), 12–13; Jimmy M. Skaggs, *Prime Cut: Livestock Raising and Meatpacking in the United States, 1607–1983* (College Station: Texas A&M University Press, 1986), 37–38.
20. The story of the origin of the name Porkopolis is from Richard G. Arms, "From

Disassembly to Assembly, Cincinnati: The Birthplace of Mass-Production," *Bulletin of the Historical and Philosophical Society of Ohio* 17 (July 1959): 195.

21. Henry W. Ellsworth, *The American Swine Breeder: A Practical Treatise on the Selection, Rearing, and Fattening of Swine* (Boston: Weeks, Jordan, 1840), 288.

22. "Isabella Bird Rides the Cars," in *With Women's Eyes: Visitors to the New World, 1775–1918*, ed. Marion Tinling (North Haven, CT: Archon Books, 1993), 95.

23. Herbert A. Kellar, "A Journey through the South in 1836: Diary of James D. Davidson," *Journal of Southern History* 1(3) (August 1935): 350–351.

24. Arnold Schrier, "A Russian Observer's Visit to 'Porkopolis,' 1857," *Cincinnati Historical Society Bulletin* 29 (Spring 1971): 33.

25. William H. Hildreth, "Mrs. Trollope in Porkopolis," *Ohio State Archaeological and Historical Quarterly* 58 (January 1949): 41–42.

26. This description of the slaughtering process is based on two 1830s accounts given in Ellsworth, *American Swine Breeder*, 288–291. Quote from Henry A. Murray, cited in Jones, *History of Agriculture in Ohio*, 130.

27. Frederick Law Olmsted, *A Journey through Texas: Or a Saddle-Trip on the Southwestern Frontier* (1857; Lincoln: University of Nebraska Press, 2004), 9.

28. Ellsworth, *American Swine Breeder*, 294–295.

29. Arms, "From Disassembly to Assembly," 200: Warren, *Tied to the Great Packing Machine*, 16.

30. Carl M. Becker, "Evolution of the Disassembly Line: The Horizontal Wheel and the Overhead Railway Loop," *Cincinnati Historical Society Bulletin* 26(3) (July 1968): 279.

31. Becker, "Evolution of the Disassembly Line," 279–281.

32. Walsh, *Rise of the Midwestern Meat Packing Industry*, 12, 45, 48; Hurt, "Pork and Porkopolis," 205, 208. For an extensive discussion of Chicago's rise as a pork packing center, see William Cronon, *Nature's Metropolis: Chicago and the Great West* (New York: Norton, 1991), chap. 5.

33. Neil Adams McNall, *An Agricultural History of the Genesee Valley, 1790–1860* (Westport, CT: Greenwood, 1952), 142, 165–166; Faragher, *Sugar Creek*, 201; Frederick Merk, *Economic History of Wisconsin during the Civil War Decade*, 2nd ed. (Madison: State Historical Society of Wisconsin, 1971), 150; Walsh, *Rise of the Midwestern Meat Packing Industry*, chap. 3.

34. Diane Wenger, "Delivering the Goods: The Country Storekeeper and Inland Commerce in the Mid-Atlantic," *Pennsylvania Magazine of History and Biography* 129(1) (January 2005): 56, 60.

35. R. Douglas Hurt, *The Ohio Frontier: Crucible of the Old Northwest, 1720–1830* (Bloomington: Indiana University Press, 1996), 215; Walsh, *Rise of the Midwestern Meat Packing Industry*, 7–811; James Stuart, *Three Years in North America*, 3rd ed. (Edinburgh: Robert Cadell, 1833), 271.
36. John C. Hudson, *Making the Corn Belt: A Geographical History of Middle-Western Agriculture* (Bloomington: Indiana University Press, 1994), 140.
37. Charles E. Brooks, *Frontier Settlement and Market Revolution: The Holland Land Purchase* (Ithaca, NY: Cornell University Press, 1996), 105.
38. Sam Bowers Hilliard, *Hog Meat and Hoecake: Food Supply in the Old South* (Carbondale: Southern Illinois University Press, 1972), 57–58; Skaggs, *Prime Cut*, 39–41.
39. S. Reynolds, "ART. 166—Remarks upon the Treatment of Swine, Read before the Farmer's Society in the Town of Florida, N.Y. November, 1828," *New York Farmer & Horticultural Repository*, November 1, 1828, 268.
40. Russell, *North America*, 79–80.
41. Harry N. Scheiber, *Ohio Canal Era: A Case Study of Government and the Economy, 1820–1861* (Athens: Ohio University Press, 1969), 219.
42. Larry Gara, "A Correspondent's View of Cincinnati in 1839," *Bulletin of the Historical and Philosophical Society of Ohio* 9 (April 1951): 135.
43. Olmsted, *Journey through Texas*, 9.
44. "American Provisions in England," *Niles' National Register*, November 2, 1844, 137; "Provision Trade with England," *Farmers Cabinet and American Herd Book*, November 15, 1843, 114; "American Provisions in England," *American Agriculturalist*, December 1, 1849, 374.
45. Hilliard, *Hog Meat and Hoecake*, 203, 215–219.
46. Russell, *North America*, 265, 290.
47. For a discussion of the perception and reality of slave theft, see Psyche A. Williams-Forson, *Building Houses out of Chicken Legs: Black Women, Food, and Power* (Chapel Hill: University of North Carolina Press, 2006), 26–31.
48. John Hebron Moore, *The Emergence of the Cotton Kingdom in the Old Southwest, Mississippi, 1770–1860* (Baton Rouge: Louisiana State University Press, 1988), 25.
49. Ulrich Bonnell Phillips, *American Negro Slavery* (New York: Appleton, 1929), 233, 237, 297; James Mallory, *"Fear God and Walk Humbly": The Agricultural Journal of James Mallory*, ed. Grady McWhiney, Warren O. Moore, Jr., and Robert F. Pace (Tuscaloosa: University of Alabama Press, 1997), 79.

NOTES TO PAGES 122–126

50. "The Editor of the Cleveland (Ohio) Herald . . . ," *Daily National Intelligencer*, December 9, 1842, column A.
51. Davis Dyer, Frederick Dalzell, and Rowena Olegario, *Rising Tide: Lessons from 165 Years of Brand Building at Procter & Gamble* (Boston: Harvard Business School Press, 2004), chap. 1.
52. "Pork and Lard Oil," *Albany Cultivator*, May 15, 1843, 54.
53. Quote from "Lard Oil," *Milwaukee Sentinel*, April 12, 1843, column G. For nicknames for hogs, see (Ohio whale) "Welsh's Lard Lamp," *Scioto Gazette*, February 9, 1843, column E; (land whale) "Lard Oil," *Mississippi Free Trader and Natchez Daily Gazette*, August 4, 1843, column A; (prairie whale) "Lard Oil," *Cleveland Daily Herald*, January 4, 1843, column C.
54. "Lard Oil Business," *Cleveland Daily Herald*, September 17, 1842, column B; "Lard Oil," *Ohio Observer*, March 9, 1843, 40; "Lard for India," *Cleveland Daily Herald*, February 27, 1851, column B; "Lard," *Ripley Bee*, February 10, 1855, column C; "Tallow and Lard," *Cleveland Daily Herald*, April 21, 1854, column B.
55. "Lard Oil," *Mississippi Free Trader and Natchez Daily Gazette*, August 4, 1843, column A.
56. Harold F. Williamson and Arnold R. Daum, *The American Petroleum Industry: The Age of Illumination, 1859–1899* (Evanston, IN: Northwestern University Press, 1959), 34–37, 43–60.
57. Richard Perren, *Taste, Trade and Technology: The Development of the International Meat Industry since 1840* (Burlington, VT: Ashgate, 2006), 40; Paul W. Gates, *Agriculture and the Civil War* (New York: Knopf, 1965), 176–177.
58. Gates, *Agriculture and the Civil War*, 176–177, 183–184; Allan G. Bogue, "Twenty Years of an Iowa Farm Business, 1860–1880," *Annals of Iowa* 35(8) (Spring 1961): 567.
59. Palmer H. Boeger, "The Great Kentucky Hog Swindle of 1864," *Journal of Southern History* 28(1) (February 1862): 59–70. For an in-depth treatment of reconciliation in Kentucky, see Anne E. Marshall, *Creating a Confederate Kentucky: The Lost Cause and Civil War Memory in a Border State* (Chapel Hill: University of North Carolina Press, 2010).
60. For an account of the Union Stockyards, see Cronon, *Nature's Metropolis*; Walsh, *Rise of the Midwestern Meat Packing Industry*, 73–74.
61. Hudson, *Making the Corn Belt*, 155.
62. Walsh, *Rise of the Midwestern Meat Packing Industry*, 84–85; Cronon, *Nature's Metropolis*, 230–232.

63. Cited in "American Pork," *Illinois Farmer*, October 1, 1863, 303.
64. John L. Gignilliat, "Pigs, Politics, and Protection: The European Boycott of American Pork, 1879–1891," *Agricultural History* 35(1) (January 1961): 4.
65. Gignilliat, "Pigs, Politics, and Protection," 3–12.
66. Gignilliat, "Pigs, Politics, and Protection," 10.
67. Clemen, *American Livestock and Meat Industry*, 321.
68. Joseph Nimmo, Jr., *The Production of Swine in the United States, and the Transportation, Consumption, and Exportation of Hog Products, with Special Reference to Interdiction of American Hog Products from France and Germany* (Washington, DC: Government Printing Office, 1884), 12–14.
69. Bingham Duncan, "Protectionism and Pork: Whitelaw Reid as Diplomat, 1889–1891," *Agricultural History* 33(4) (October 1959): 195; Gignilliat, "Pigs, Politics, and Protection," 11–12.
70. Thomas Shaw, "The Bacon Hog," *Prairie Farmer*, January 7, 1899, 3; "Class of Hogs Wanted for Bacon," *Prairie Farmer*, December 18, 1897, 2.
71. I. N. Cowdry, "Streaks of Lean," in *Special Report of the Indiana State Board of Agriculture on the Hog* (Indianapolis: Wm. B. Burford, 1900), 152–153.
72. "The Class of Hogs Packers Want," *Prairie Farmer*, January 15, 1898, 3.
73. "Lean Hogs Must Be Furnished," *Prairie Farmer*, July 25, 1896, 1.
74. Germantown Telegraph cited in "Fattening Hogs," *Massachusetts Ploughman and the New England Journal of Agriculture*, October 20, 1877, 1.
75. A. B. Genung, "Agriculture in the World War Period," in *Farmers in a Changing World: Yearbook of Agriculture, 1940* (Washington, DC: Government Printing Office, 1940), 286; Clemen, *American Livestock and Meat Industry*, 286, 304; Perren, *Taste, Trade and Technology*, 94–95.
76. Benjamin H. Hibbard, *Effects of the Great War upon Agriculture in the United States and Great Britain* (New York: Oxford University Press, 1919), 63; Frank M. Surface, *American Pork Production in the World War* (Chicago: A.W. Shaw, 1926), 18.
77. Robert D. Cuff, "The Limits of Voluntarism: The Pork-Packing Agreement of 1917–1919," *Agricultural History* 53(4) (October 1979): 727–747.
78. P. H. Rolfs, "Our Nation Is Calling for More Hogs at High Prices," Extension Circular No. 1 (Gainesville: University of Florida Division of Agricultural Extension and US Department of Agriculture, n.d.), University of Florida Digital Collections. http://ufdc.ufl.edu/UF00084625/00001/1?search=nation+%3dcalling+%3dhogs.

79. "How Many Sows to Breed," *Wallaces' Farmer and Iowa Homestead*, June 8, 1917, 5.
80. Walter T. Borg, "Food Administration Experience with Hogs, 1917–1919," *Journal of Farm Economics* 25(2) (May 1934): 456; Gary Dean Best, "Food Relief as a Price Support: Hoover and American Pork, January–March 1919," *Agricultural History* 45(2) (April 1971): 80, 82–84.
81. George T. Blakey, "Ham That Never Was: The 1933 Emergency Hog Slaughter," *Historian* 31(1) (November 1967):, 42.
82. Blakey, "Ham That Never Was," 42.
83. Blakey, "Ham That Never Was," 42–46; Theodore Saloutos, *The American Farmer and the New Deal* (Ames: Iowa State University Press, 1982), 70–73.
84. C. Roger Lambert, "Slaughter of the Innocents: The Public Protests AAA Killing of Little Pigs," *Midwest Quarterly* 14(3) (Spring 1973): 247–254, 253; C. Roger Lambert, "'Slaughter of the Innocents' in Oklahoma: The Emergency Hog Slaughter of 1933," *Red River Valley Historical Review* 7(4) (1982): 42–49; H. Roger Grant and L. Edward Purcell, eds., *Years of Struggle: The Farm Diary of Elmer G. Powers* (DeKalb: Northern Illinois University Press, 1995), 64.
85. Grant and Purcell, *Years of Struggle*, 78, 80, 122–123; Perren, *Taste, Trade and Technology*, 110; Geofrey Shepherd, J. C. Purcell, and L. V. Manderscheid, "Economic Analysis of Trends in Beef Cattle and Hog Prices," *Research Bulletin* 405 (Ames: Agricultural Experiment Station, Iowa State College, January 1954), 730.
86. Walter W. Wilcox, *The Farmer in the Second World War* (Ames: Iowa State College Press, 1947), 163–166.
87. Wilcox, *Farmer in the Second World War*, 163–166; *U.S. Census of Agriculture: 1945*, vol. 2, pt. 7 (Washington, DC: Government Printing Office, 1947), 347.
88. "Fewer Pigs Coming," *Wallaces' Farmer and Iowa Homestead*, January 6, 1945, 1.
89. Richard Dougherty, *In Quest of Quality: Hormel's First 75 Years* (Austin, MN: George Hormel and Company, 1966), 197; Jenny Leigh Smith, "Tushonka: Cultivating Soviet Postwar Taste," *M/C Journal* 13(5) (2010): 178–181.
90. Shepherd, Purcell, and Manderscheid, "Economic Analysis of Trends," 739–741.
91. Rolland "Pig" Paul, "Pork Organizations, 1940–1975," in *The Pork Story: Legend and Legacy*, ed. Rolland "Pig" Paul, J. Marvin Garner, and Orville Sweet (Kansas City, MO: Lowell Press, 1991), 52–53.
92. Paul, "Pork Organizations," 58.

93. Erin Borror, "NAFTA Turns 20: A Closer Look at the North American Red Meat Trade" (US Meat Export Federation, n.d.), http://www.usmef.org/nafta-turns-20-a-closer-look-at-north-american-red-meat-trade.
94. "Pork That Is Pinker and Purer Appeals to the Japanese," *New York Times*, October 1, 1989, F17.
95. William A. Amponsah, Xiang Dong Qin, and Xuehua Peng, "China as a Potential Market for U.S. Pork Exports," *Review of Agricultural Economics* 25(1) (2003): 259–260.
96. "U.S. Pork 2016 Export Data Shows Impressive Progress" (National Pork Board, Pork Checkoff, 2016), http://www.pork.org/u-s-pork-2016-export-data-shows-impressive-progress/; Malinda Geisler, revised by Kelly Collins, "Pork International Markets Profile" (Ag Marketing Resource Center, 2017), http://www.agmrc.org/commodities-products/livestock/pork/pork-international-markets-profile/; S.-H. Oh and M. T. See, "Pork Preference for Consumers in China, Japan and South Korea," *Asian-Australian Journal of Animal Science* 25(1) (January 2012): 145.
97. David L. Ortega, "Assessing the Opportunities for U.S. Pork in China" (Purdue Agricultural Economics Report, November 2008), http://www.agecon.purdue.edu/extension/pubs/paer/2008/november/wang.asp; Michael J. de la Merced and David Barboza, "Needing Pork, China Is to Buy a U.S. Supplier," *New York Times*, May 30, 2013, A1.

## CHAPTER 6

1. *Ohio Farmer* report cited in *American Farmer*, April 1857, 323; *Daily National Intelligencer*, November 5, 1857, 123; "The Cholera among Hogs," *Illinois Farmer*, April 1857, 84; "The Hog Cholera in New Jersey," *New York Herald*, May 15, 1857, 5; Princeton quote from "Hog Cholera—The Remedy," *American Farmer*, September 1858, 73; Cartmell quote from Donald L. Winters, *Tennessee Farming, Tennessee Farmers: Antebellum Agriculture in the Upper South* (Knoxville: University of Tennessee Press, 1994), 71.
2. G. H. Wise, "Hog Cholera and Its Eradication: A Review of U.S. Experience," US Department of Agriculture, Animal and Plant Health Inspection Service, APHIS 91-95 (Washington, DC: Government Printing Office, September 1981), 1–2; C. G. Cole, R. R. Henley, C. N. Dale, L. O. Mott, J. P. Torrey, and M. R. Zinober, "History of Hog Cholera Research in the U.S. Department of Agriculture," Agriculture Information Bulletin No. 241 (Washington, DC: US Department of Agriculture, Agricultural Research Service, 1962), 2.

3. Henry W. Ellsworth, *The American Swine Breeder: A Practical Treatise on the Selection, Rearing, and Fattening of Swine* (Boston: Weeks, Jordan, 1840), 274.
4. Robert Leslie Jones, *History of Agriculture in Ohio to 1880* (Kent: Kent State University Press, 1983), 123, 350n67.
5. Alan L. Olmstead and Paul W. Rhode, *Arresting Contagion: Science, Policy, and Conflicts over Animal Disease Control* (Cambridge, MA: Harvard University Press, 2015), 140.
6. Ellsworth, *American Swine Breeder*, 274–275.
7. "Hog Cholera—Kentucky Premium," *Illinois Farmer*, June 1860, 91; "Hog Cholera—The Remedy," *American Farmer*, September 1858, 73.
8. Edwin Snow, "Hog Cholera," US Patent Office Report for 1861 (Washington, DC: Government Printing Office, 1862), 152–153.
9. Winters, *Tennessee Farming, Tennessee Farmers*, 71.
10. Grady McWhiney, Warren O. Moore, Jr., and Robert F. Pace, eds., *"Fear God and Walk Humbly": The Agricultural Journal of James Mallory* (Tuscaloosa: University of Alabama Press, 1997), 468.
11. "Losses of Swine," *Milwaukee Daily Sentinel*, May 30, 1877, 5.
12. N. H. Paaren, "Veterinary Sanitary Reform," *Prairie Farmer*, March 9, 1878, 5.
13. Virginia E. McCormick, ed., *Farm Wife: A Self-Portrait, 1886–1896* (Ames: Iowa State University Press, 1990), 55–56.
14. "Apples and Hog Cholera," *Prairie Farmer*, July 24, 1875, 2; "The Fall Management of Pigs, *Wallaces' Farmer and Iowa Homestead*, September 2, 1898, 7; "Notes from Correspondents," *Prairie Farmer*, December 21, 1872, 8; F. D. Coburn, *Swine Husbandry* (New York: Orange Judd, 1900), 130; "Hog Cholera," *Prairie Farmer*, November 13, 1875, 5.
15. "Hog Cholera," *Prairie Farmer*, November 16, 1872, 5.
16. O. H. V. Stalheim, *The Winning of Animal Health: 100 Years of Veterinary Medicine* (Ames: Iowa State University Press, 1994), chap. 5; G. H. Wise, "Hog Cholera and Its Eradication," 2.
17. Cole et al., "History of Hog Cholera Research," 6.
18. Richard A. Overfield, "Hog Cholera, Texas Fever, and Frank S. Billings: An Episode in Nebraska Veterinary Science," *Nebraska History* 57(1) (March 1976): 99–128.
19. Overfield, "Hog Cholera," 110–112.
20. Frank S. Billings, *Inoculation a Preventive of Swine Plague, with the Demonstration That the Administration of the Agricultural Department Is a Public Scandal: An Exposure* (Lincoln, NE: State Journal Company, 1892), 5–6.

21. "The Hog Cholera Problem," *Wallaces' Farmer and Iowa Homestead*, February 16, 1900, 20.
22. Cole et al., "History of Hog Cholera Research," 8–9.
23. "Supplement to History of Hog Cholera Research in the U.S. Department of Agriculture," Supplement to Agriculture Information Bulletin No. 241 (Washington, DC: Government Printing Office, March 1962), 1; Cole et al., "History of Hog Cholera Research," 9–11; Wise, "Hog Cholera and Its Eradication," 2–3.
24. Wise, "Hog Cholera and Its Eradication," 4; Brian W. Beltman, *Dutch Farmer in the Missouri Valley: The Life and Letters of Ulbe Uringa, 1866–1950* (Urbana: University of Illinois Press, 1996), 90; "Deer Island Hogs Doomed by Cholera," *Boston Daily Globe*, February 28, 1918, 12.
25. J. P. Torrey, "Hog Cholera," in *Yearbook of Agriculture, 1956* (Washington, DC: Government Printing Office, 1956), 359–360; Cole et al., "History of Hog Cholera Research," 12–19; Paul De Kruif, "Dorset Digs a Grave for Hog Cholera," *Country Gentleman*, October 1926, 80–81.
26. Cole et al., "History of Hog Cholera Research," 16–20.
27. Proceedings of the Conference of Federal and State Officers Held in Ames, Iowa, May 30, 1908, Iowa State University Library, Special Collections Department, R. A. (Raymond Allen) Packer Papers, RS 14/6/11, series 4, box 7, folder 4; M. Dorset, "Hog Cholera," US Department of Agriculture Farmers' Bulletin No. 379 (Washington, DC: Government Printing Office, 1909); "A New Era in Pork Production," *Berkshire World and Corn Belt Stockman*, December 1, 1910, 3.
28. Cole et al., "History of Hog Cholera Research," 16–20.
29. G. R. White, "Hog Cholera and Serum Treatment," in *Farmers' Bulletin* (Nashville: Department of Agriculture, State of Tennessee, 1914), 2.
30. Robert H. Ferrell, *Harry S. Truman: His Life on the Family Farms* (Worland, WY: High Plains, 1991), 55–56.
31. Ferrell, *Harry S. Truman*, 55.
32. E. F. Lowry, *Don't Vaccinate Your Hogs: Hog Cholera and Other Hog Sickness and the Failure of the Serum Treatment* (Ottumwa, IA: Commercial Printing, 1913), 23–32; Wise, "Hog Cholera and Its Eradication," 5.
33. J. W. Connaway, "Stamping Out Hog Cholera," *Missouri Farmer*, 1915, 7.
34. "Another Hog Cholera 'Cure' Fraud," *Berkshire World and Corn Belt Stockman*, August 1, 1914, 1; C. H. Stange and C. G. Cole, "Facts about So-Called Hog Cholera Cures and Specifics," Circular No. 25, (Agricultural Experiment Station, Iowa State College of Agriculture and Mechanic Arts, Veterinary

Section, August 1915), n.p.; "Hog Cholera Dopes," *Wallaces' Farmer and Iowa Homestead*, May 24, 1918, 6; Charles F. Lynch, *Diseases of Swine, with Particular Reference to Hog Cholera* (Philadelphia: W.B. Saunders, 1914), 318.

35. Lynch, *Diseases of Swine*, 315–316.
36. C. A. Lueder, *Hog Cholera: Its Prevention and Control* (Charleston: West Virginia Experiment Station, June 1913), 8–10, 12–15.
37. Wise, "Hog Cholera and Its Eradication," 7; "Supplement to History of Hog Cholera Research," 83.
38. Wise, "Hog Cholera and Its Eradication," 5, 7; T. P. White, "Diseases, Ailments, and Abnormal Conditions of Swine," US Department of Agriculture Farmers' Bulletin No. 1244 (Washington, DC: Government Printing Office, 1923; revised 1931, 1935), 1.
39. Wise, "Hog Cholera and Its Eradication," 8.
40. Wise, "Hog Cholera and Its Eradication," 9–11, 13.
41. Wise, "Hog Cholera and Its Eradication," 13.
42. The four-phase program is elaborated in Wise, "Hog Cholera and Its Eradication," 16–17.
43. Animal and Plant Health Inspection Service, US Department of Agriculture, "Cooperative State-Federal Hog Cholera Eradication Program Progress Report 1971," APHIS 91-9 (Washington, DC: Government Printing Office, November 1972), 1–3.
44. Wise, "Hog Cholera and Its Eradication," 17.
45. Wise, "Hog Cholera and Its Eradication," 17–20.
46. Richard P. Horwitz, *Hog Ties: Pigs, Manure, and Mortality in American Culture* (New York: St. Martin's, 1998), 15–17.
47. Joe Vansickle, "Facing Chronic Disease," *National Hog Farmer*, January 15, 2007, http://nationalhogfarmer.com/mag/farming_facing_chronic_disease.
48. "Top Ten Ways to Boost Biosecurity," *National Hog Farmer*, June 15, 2007, http://nationalhogfarmer.com/mag/farming_boost_biosecurity.
49. Alex Blanchette, "Herding Species: Biosecurity, Posthuman Labor, and the American Industrial Pig," *Cultural Anthropology* 30(4) (2015): 652–653.
50. Kent Schwartz and Rodger Main, "Porcine Epidemic Diarrhea (PED) Virus: FAQ and Survival Tips," *National Hog Farmer*, July 29, 2013, http://nationalhogfarmer.com/health/porcine-epidemic-diarrhea-virus-faq-and-survival-tips.
51. Changhee Lee, "Porcine Epidemic Diarrhea Virus: An Emerging and Re-emerging Epizootic Swine Virus," *Virology Journal* 12(193) (2015): 1–16.
52. Reed Karaim, "Virus Kills Millions of American Pigs, Pushing Up Pork

Prices," *National Geographic*, May 1, 2014, http://news.nationalgeographic.com/news/2014/05/140501-pigs-virus-meat-prices-food-science-health/.

53. Betsy Freese, "Pork Powerhouses 2013: Disease Hits, Growth Continues," Agriculture.com, September 30, 2013, http://www.agriculture.com/livestock/hogs/pk-powerhouses-2013-disease-hits-growth_283-ar34203?print.

54. "Disease Outbreaks in Hog Industry Continue," *Successful Farming*, April 2004, 49.

## CHAPTER 7

1. Joe Gray Taylor, *Eating, Drinking, and Visiting in the South: An Informal History* (Baton Rouge: Louisiana State University Press, 1982), 111–112, 113; William Hampton Adams and Steven D. Smith, "Historical Perspectives on Black Tenant Farmer Material Culture: The Henry C. Long General Store Ledger at Waverly Plantation, Mississippi," in *The Archaeology of Slavery and Plantation Life*, ed. Theresa A. Singleton (Walnut Creek, CA: Left Coast Press, 2009); 320–323.

2. W. O. Atwater and Charles D. Woods, "Dietary Studies with Reference to the Food of the Negro in Alabama in 1895 and 1896," US Department of Agriculture Office of Experiment Stations Bulletin No. 38 (Washington, DC: Government Printing Office, 1897), 21. Dietary studies proliferated during the late nineteenth and early twentieth centuries as researchers at state agricultural experiment stations, armed with new concepts such as the calorie and vitamins, began to study the diets of various socioeconomic and ethnic groups across the country.

3. Edward King, *The Great South: A Record of Journeys in Louisiana, Texas, the Indian Territory, Missouri, Arkansas, Mississippi, Alabama, Georgia, Florida, South Carolina, North Carolina, Kentucky, Tennessee, Virginia, West Virginia, and Maryland* (Hartford, CT: American Publishing, 1875), 791.

4. Amy Feely Morsman, *The Big House after Slavery: Virginia Plantation Families and Their Postbellum Domestic Experience* (Charlottesville: University of Virginia Press, 2010), 43.

5. Quoted in Reginald Horsman, *Feast or Famine: Food and Drink in American Westward Expansion* (Columbia: University of Missouri Press, 2008), 304.

6. H. Arnold Barton, ed., *Letters from the Promised Land: Swedes in America, 1840–1914* (Minneapolis: University of Minnesota Press, 1975), 144, 151.

7. William D. Rowley, *Reclaiming the Arid West: The Career of Francis G. Newlands* (Bloomington: Indiana University Press, 1996), 54.

8. H. B. Gibson, S. Calvert, and D. W. May, "Dietary Studies at the University of Missouri in 1895," US Department of Agriculture Office of Experiment Stations Bulletin No. 31 (Washington, DC: Government Printing Office, 1896), 8.
9. Allan G. Bogue, "Twenty Years of an Iowa Farm Business, 1860–1880," *Annals of Iowa* 35(8) (Spring 1961): 571; Virginia E. McCormick, ed., *Farm Wife: A Self-Portrait, 1886–1896* (Ames: Iowa State University Press, 1990), 85–86; "The Yearly Round of Isaac N. Carr (1897)" (unpublished manuscript, Living History Farms, Urbandale, IA, n.d.); Merrill Jarchow, "Life on a Jones County Farm, 1873–1912," *Iowa Journal of History* 49(4) (October 1951): 320.
10. Harvey Levenstein, *Revolution at the Table: The Transformation of the American Diet* (Berkeley: University of California Press, 2003), 21; Gibson, Calvert, and May, "Dietary Studies," 8; Richard J. Hooker, *Food and Drink in America: A History* (Indianapolis, IN: Bobbs-Merrill, 1981), 222.
11. E. F. Haskell, *The Housekeeper's Encyclopedia of Useful Information for the Housekeeper in All Branches of Cooking and Domestic Economy* (New York: Appleton, 1864), iv–v; Hooker, *Food and Drink in America*, 217.
12. Morsman, *Big House after Slavery*, 69–70.
13. Levenstein, *Revolution at the Table*, 21, 22.
14. Hooker, *Food and Drink in America*, 222; Fred Shannon, "The Status of the Midwestern Farmer in 1900," *Mississippi Valley Historical Review* 37(3) (December 1950): 497.
15. Susan Williams, *Food in the United States, 1820s–1890* (Westport, CT: Greenwood, 2006), 108–109.
16. Gibson, Calvert, and May, "Dietary Studies," 8.
17. Hooker, *Food and Drink in America*, 237–238; Walter D. Kamphoefner, Wolfgang Helbich, and Ulrike Sommer, *News from the Land of Freedom: German Immigrants Write Home* (Ithaca, NY: Cornell University Press, 1991), 469.
18. For both sides of the scrapple issue, see "More about 'Scrapple,'" *New York Times*, February 2, 1872, 2 and "Trying Scrapple," *New York Times*, February 6, 1872, 2.
19. "How Poor People Live: Talk with a Working Man-What He Makes, What He Spends, and What He Saves," *New York Times*, January 28, 1872, 5.
20. Robert T. Dirks and Nancy Duran, "African American Dietary Patterns at the Beginning of the Twentieth Century," *Journal of Nutrition* 131(7) (July 2001): 1887; Mark Warner, "Ham Hocks on Your Cornflakes," *Archaeology* 54(6) (November–December 2001): 202–206.
21. Arthur Goss, "Dietary Studies in New Mexico," US Department of Agriculture

Office of Experiment Stations Bulletin No. 40 (Washington, DC: Government Printing Office, 1897), 6; Jarchow, "Life on a Jones County Farm," 319–320; McCormick, *Farm Wife*, 86.

22. Cottolene advertisements, *Bangor Daily Whig and Courier*, June 28, 1892; *North American* (Philadelphia), September 20, 1897, 6; *Macon Telegraph*, November 20, 1897, 2; December 1, 1897, 3; and December 3, 1897, 6.

23. "Our Friend the Hog," *Wallaces' Farmer and Iowa Homestead*, June 7, 1907, 14.

24. Skaggs, *Prime Cut*, 131–132; Amy Bentley, *Eating for Victory: Food Rationing and the Politics of Domesticity* (Urbana: University of Illinois Press, 1998), 92; Roger Horowitz, *Putting Meat on the American Table: Taste, Technology, Transformation* (Baltimore: Johns Hopkins University Press, 2006), 44.

25. Earl Shaw, "Swine Production in the Corn Belt of the United States," *Economic Geography* 12(4) (October 1936): 368–369.

26. Horowitz, *Putting Meat on the American Table*, 55, 58–60, 62.

27. Rupert B. Vance, *Human Geography of the South: A Study in Regional Resources and Human Adequacy* (Chapel Hill: University of North Carolina Press, 1932), 423–424; Susan J. Mathews, "Food Habits of Georgia Rural People," Georgia Experiment Station Bulletin No. 159 (Experiment, GA: Georgia Experiment Station, May 1929), 28; Clifton L. Paisley, "Van Brunt's Store, Iamonia, Florida, 1902–1911," *Florida Historical Quarterly*, 48(4) (April 1970): 359.

28. Arthur F. Raper, *Preface to Peasantry: A Tale of Two Black Belt Counties*, repr. ed. (Columbia: University of South Carolina Press, 2005), 42–43; Frederick Douglass Opie, *Hog and Hominy: Soul Food from Africa to America* (New York: Columbia University Press, 2008), 60–61.

29. Jimmye Hillman, *Hogs, Mules, and Yellow Dogs: Growing Up on a Mississippi Subsistence Farm* (Tucson: University of Arizona Press, 2012), 180.

30. Katherine Leonard Turner, *How the Other Half Ate: A History of Working-Class Meals at the Turn of the Century* (Berkeley: University of California Press, 2014), 108–110; Taylor, *Eating, Drinking, and Visiting*, 112.

31. Robert F. Moss, *Barbecue: The History of an American Institution* (Tuscaloosa: University of Alabama Press, 2010), 153–159.

32. Cited in Jessica B. Harris, *High on the Hog: A Culinary Journey from Africa to America* (New York: Bloomsbury, 2011), 179, 176–183; Megan J. Elias, *Food in the United States, 1890–1945* (Santa Barbara, CA: Greenwood, 2009), 15; Adrian Miller, *Soul Food: The Surprising Story of an American Cuisine, One Plate at a Time* (Chapel Hill: University of North Carolina Press, 2013), 155.

33. John W. Bennett, Harvey L. Smith, and Herbert Passin, "Food and Culture in

Southern Illinois—A Preliminary Report," *American Sociological Review* 7(5) (October 1942): 65.

34. Carl Hamilton, *In No Time at All* (Ames: Iowa State University Press, 1974), 46.
35. Peter A. Speek, *A Stake in the Land* (New York: Harper, 1921), 208.
36. Robert V. Tauxe and Emilio J. Esteban, "Advances in Food Safety to Prevent Foodborne Diseases in the United States," in *Silent Victories: The History and Practice of Public Health in Twentieth Century America*, ed. John W. Ward and Christian Warren (Oxford: Oxford University Press, 2007), 27.
37. Benjamin Schwartz, "Trichinosis: A Disease Caused by Eating Raw Pork," US Department of Agriculture Leaflet No. 34 (Washington, DC: Government Printing Office, May 1929, slightly revised February 1941), 1.
38. *Meat for the Millions: Report of the New York State Trichinosis Commission* (Albany, NY: Fort Orange Press, 1941), 14, 63–66.
39. Caroline L. Hunt, "Good Portions in the Diet," US Department of Agriculture Farmers' Bulletin No. 1313 (Washington, DC: Government Printing Office, March 1923), 3.
40. Horowitz, *Putting Meat on the American Table*, 65–66; James Shellenberger, "Meat on the March," *DuPont Magazine*, February 1941, 11.
41. Richard Dougherty, *In Quest of Quality: Hormel's First 75 Years* (Austin, MN: George A. Hormel and Company, 1966), 163.
42. Bentley, *Eating for Victory*, 71, 174; Dougherty, *In Quest of Quality*, 197–199.
43. "Family Food Consumption in the United States, Spring 1942," US Department of Agriculture Miscellaneous Publication No. 550 (Washington, DC: Government Printing Office, 1944), 14.
44. Merrill K. Bennett and Rosamond H. Peirce, "Change in the American National Diet, 1879–1959," *Food Research Institute Studies* 2(2) (1961): 104–107.
45. Harvey Levenstein, *Paradox of Plenty: A Social History of Eating in Modern America* (Berkeley: University of California Press, 2003), 207–209.
46. Lawrence A. Duewer, Kenneth R. Krause, and Kenneth E. Nelson, "U.S. Poultry and Red Meat Consumption, Prices, Spreads, and Margins," Agriculture Information Bulletin No. 684 (Washington, DC: USDA, Economic Research Service, October, 1993), 1–3; William Boyd, "Making Meat: Science, Technology, and American Poultry Production," *Technology and Culture* 42 (October 2001): 631–664; Roger Horowitz, "Making the Chicken of Tomorrow: Reworking Poultry as Commodities and as Creatures, 1945–1990," in *Industrializing Organisms: Introducing Evolutionary History*, ed. Susan R. Schrepfer and

Philip Scranton (New York: Routledge, 2004); Horowitz, *Putting Meat on the American Table*, chap. 5.

47. "Farms Retooling for New Model Pig," *Nation's Business*, October 1955, 44.
48. Bennett and Peirce, "Change in the American National Diet," 104–107; David L. Call and Ann MacPherson Sánchez, "Trends in Fat Disappearance in the United States, 1909–1965," *Journal of Nutrition* 93 (1967): 17.
49. "We've Got to Do a Major Job Remodeling Hogs," *Country Gentleman*, March 1953, 204.
50. "Farms Retooling for New Model Pig," *Nation's Business*, October 1955, 44.
51. Dik Twedt, "General Acceptance of Pork," in *The Pork Industry: Problems and Progress*, ed. David G. Topel (Ames: Iowa State University Press, 1968), 3.
52. University of Illinois Agricultural Extension, "Pork People Like," Prelinger Archive, 1956, https://archive.org/details/PorkPeop1956; Iowa State College, "The Pig and the Public," Iowa State University Library, Special Collections Department, https://www.youtube.com/watch?v=G-oG7yL4_mA&t=4s.
53. J. L. Anderson, "Lard to Lean: Making the Meat-Type Hog in Post–World War II America," in *Food Chains: From Farmyard to Shopping Cart*, ed. Warren Belasco and Roger Horowitz (Philadelphia: University of Pennsylvania Press, 2009); William Colgan Page, *Leaner Pork for a Healthier America: Looking Back on the Northeast Iowa Swine Testing Station* (Iowa Department of Transportation in cooperation with the Federal Highway Administration and the State Historical Society of Iowa, 2000); Don Muhm, ed., *Iowa Pork and People: A History of Iowa's Pork Producers* (Clive, IA: Iowa Pork Producers, 1995), chap. 10; K. J. Drewry, "How to Make Backfat Determinations," Paper No. 156 (West Lafayette, IN: Purdue University Extension, 1977).
54. Committee on Technological Options to Improve the Nutritional Attributes of Animal Products, Board on Agriculture, National Research Council, *Designing Foods: Animal Product Options in the Marketplace* (Washington, DC: National Academies Press, 1988), 81; Anderson, "Lard to Lean," 39.
55. *The New Pork: An Exciting Taste Treat* (Fred A. Niles Communications Centers, Pork Industry Committee of the National Livestock and Meat Board, n.d.), https://archive.org/details/NewPorkAnExcitingTaste.
56. Ralston Purina Company, *American Heirloom Pork Cookbook* (New York: McGraw-Hill, 1971), 5.
57. Anderson, "Lard to Lean," 40–41.
58. "The Meat Type Hog and Consumer's Spendable Income," Mississippi State

University Archives, Mitchell Memorial Library, W. L. Richmond Papers, Collection A96–19, box 5.

59. V. James Rhodes, H. D. Naumann, Elmer R. Kiehl, and E. A. Jaenks, "Consumer Acceptance of Lean Pork," *Journal of Farm Economics* 42(1) (February 1960): 18–34.
60. Anderson, "Lard to Lean," 42–44.
61. David Huinker, "Thoughts about Hogs . . . and Change," in Muhm, *Iowa Pork and People*, 180; Edward Behr, "The Lost Taste of Pork: Finding a Place for the Iowa Family Farm," *Art of Eating*, no. 51 (Summer 1999): 7.
62. Nancy Shute, "Building a Better Pig," *U.S. News & World Report* 139(6) August 15, 2005.
63. Behr, "Lost Taste of Pork," 7.
64. Nathanael Johnson, "Swine of the Times: The Making of the Modern Pig," *Harper's Magazine*, May 2006, 47–56.
65. Steven Striffler, *Chicken: The Dangerous Transformation of America's Favorite Food* (New Haven, CT: Yale University Press, 2007), 17–19, 26–29.
66. M. S. Honeyman, R. S. Pirog, G. H. Huber, P. J. Lammers, and J. R. Hermann, "The United States Pork Niche Market Phenomenon," *Journal of Animal Science* 84 (2006): 2269–2275.
67. R. W. Apple, Jr., "A Prince of Pork: In Seattle, Recreating the Perfect Ham," *New York Times*, May 17, 2006; Christine Muhlke, "Aging Gracefully," *New York Times*, February 1, 2009; Lynne Rosetto Kasper, "Is This Prosciutto Heaven? No, It's Iowa," *Splendid Table*, January 28, 2014, http://www.splendidtable.org/story/is-this-prosciutto-heaven-no-its-iowa.
68. Burkhard Bilger, "True Grits," *New Yorker*, October 31, 2011, 40–41.
69. "Reinventing the Pig," *Des Moines Register*, September 16, 2012, 4E.
70. Michael S. Sanders, "An Old Breed of Hungarian Pig Is Back in Favor," *New York Times*, April 1, 2009.
71. Shute, "Building a Better Pig."
72. Christopher G. Davis and Biing-Hwan Lin, "Factors Affecting U.S. Pork Consumption," Electronic Outlook Report from the Economic Research Service LDP-M-130-1 (Washington, DC: US Department of Agriculture, May 2005).
73. Davis and Lin, "Factors Affecting U.S. Pork Consumption"; Eunice Romero-Gwynn et al., "Dietary Change among Latinos of Mexican Descent in California," *California Agriculture* 46(4) (July–August 1982): 12.
74. US Department of Agriculture, Office of Communications, "Profiling Food

Consumption in America," in *Agriculture Fact Book, 2001–2002* (Washington, DC: Government Printing Office, March 2003), 17.

75. Matt Lee and Ted Lee, "Light, Fluffy: Believe It, It's Not Butter," *New York Times*, October 11, 2000; Melissa Clark, "Heaven in a Pie Pan: The Perfect Crust," *New York Times*, November 15, 2006.

76. Scott Holzer, "The Modernization of Southern Foodways: Rural Immigration to the Urban South during World War II," *Food and Foodways* 6(2) (1996): 102; Charles F. Kovaicik, "Eating Out in South Carolina's Cities," and Michael O. Roark, "Fast Foods: American Food Regions," both in *The Taste of American Place: A Reader on Regional and Ethnic Foods*, ed. Barbara G. Shortridge and James R. Shortridge (Lanham, MD: Rowman & Littlefield, 1998).

77. Harris, *High on the Hog*, 199–201.

78. Rosetta E. Ross, *Witnessing and Testifying: Black Women, Religion, and Civil Rights* (Minneapolis, MN: Augsburg Fortress, 2003), 109–110; Chana Kai Lee, *For Freedom's Sake: The Life of Fannie Lou Hamer* (Urbana: University of Illinois Press, 1999), 148.

79. Harris, *High on the Hog*, 206–208.

80. Quoted in Miller, *Soul Food*, 44–45.

81. Harris, *High on the Hog*, 209–211.

82. Tony Larry Whitehead, "Sociocultural Dynamics and Food Habits in a Southern Community," in *Food in the Social Order: Studies of Food and Festivities in Three American Communities. Mary Douglas Collected Works*, vol. 9 (London: Routledge, 1984), 117–118.

83. N. W. Jerome, "Northern Urbanization and Food Consumption Patterns of Southern-Born Negroes," *American Journal of Clinical Nutrition* 22(12) (December 1969): 1667–1669.

84. Drucilla Byars, "Traditional African American Foods and African Americans," *Agriculture and Human Values* 13(1) (Summer 1996): 74–78.

85. Harris, *High on the Hog*, 225–226.

86. Renee McCoy, "African American Elders, Cultural Traditions, and the Family Reunion," *Generations: Journal of the American Society on Aging* 35(3) (Fall 2011): 17–18.

87. *Quick Facts: The Pork Industry at a Glance* (Des Moines, IA: National Pork Board, 2009–2011), 19–28.

88. Brenda Junkin, "Soldiering On: Spam and World War II," *Cleveland.com*, September 14, 2007, http://blog.cleveland.com/pdextra/2007/09/soldiering_on_spam_and_world_w.html.

89. George H. Lewis, "From Minnesota Fat to Seoul Food: Spam in America and the Pacific Rim," *Journal of Popular Culture* 34(2) (Fall 2000): 90.
90. David Sax, "The Bacon Boom Was Not an Accident," *Businessweek*, October 6, 2014.
91. Bruce Horovitz, "Carl's Jr., Hardee's New Burger Piles on the Bacon," *USA Today*, March 19, 2014, http://www.usatoday.com/story/money/business/2014/03/19/carls-jr-hardees-bacon-bacon-strips/6600167/.
92. Megan Pellegrini, "The 2012 Bacon Report," *National Provisioner*, July 31, 2012; Marc Lallanilla, "Does America Need a Bacon Intervention?," *Livescience*, February 11, 2014, http://www.livescience.com/43290-bacon-products-foods-obsession.html.
93. Bruce Horovitz, "Little Caesars Tries Bacon-Wrapped Crust Pizza," *USA Today*, February 18, 2015, http://www.usatoday.com/story/money/2015/02/18/little-caesars-fast-food-pizza-restaurants/23565411/.
94. See http://blueribbonbaconfestival.com and http://baconfestchicago.com.
95. "'United States of Bacon' TV Show on Destination America to Premier December 30," *Huffington Post*, December 5, 2012, http://www.huffingtonpost.com/2012/12/05/united-states-of-bacon-tv_n_2243675.html.
96. Lallanilla, "Does America Need a Bacon Intervention?"

## CHAPTER 8

1. "Building Better Hogs," *Wallaces Farmer*, September 1, 1962.
2. Thomas P. Janes, *A Manual on the Hog*, Circular No. 40 (Atlanta: Jas. P. Harrison, 1877), 9–10.
3. *Special Report of the Indiana State Board of Agriculture on the Hog* (Indianapolis, IN: Wm. B. Burford, 1900), 108–110.
4. Mathew R. Walpole, "The Closing of the Open Range in Watauga County, N.C.," *Appalachian Journal* 16(4) (Summer 1989): 322.
5. Steven Hahn, *The Roots of Southern Populism: Yeoman Farmers and the Transformation of the Georgia Upcountry, 1850–1890* (Oxford: Oxford University Press, 1983), chap. 7; R. Ben Brown, "Free Men and Free Pigs: Closing the Southern Range and the American Property Tradition," *Radical History Review* 108 (Fall 2010): 133.
6. Forrest McDonald and Grady McWhiney, "From Self-Sufficiency to Peonage: An Interpretation," *American Historical Review* 85(5) (December 1980): 1114.
7. W. Arthur Ayres, "Where and How to Select Breeding Animals," in *Special Report of the Indiana State Board of Agriculture on the Hog*, 87.

8. D. Trott, "Raise Only the Best Hogs," *Prairie Farmer*, June 17, 1899, 1.
9. Jacob Biggle, *Biggle Swine Book: Much Old and More New Hog Knowledge, Arranged in Alternate Streaks of Fat and Lean* (Philadelphia: Wilmer Atkinson, 1899), 7–8.
10. Joseph Harris, *Harris on the Pig: Practical Hints for the Pig Farmer* (New York: Orange Judd, 1870), 18. See also S. M. Shepard, *The Hog in America, Past and Present* (Indianapolis: Swine Breeders' Journal, 1886), 139.
11. S. M. Tracy, "Hog Raising in the South," US Department of Agriculture Farmers' Bulletin No. 100 (Washington, DC: Government Printing Office, 1899), 6.
12. Tracy, "Hog Raising in the South," 12.
13. Shepard, *Hog in America*, 80–81, 78.
14. Harris, *Harris on the Pig*, 251; Tracy, "Hog Raising in the South," 10.
15. Wenger's New Breeding Crate advertisement, *Wallaces' Farmer and Iowa Homestead*, November 30, 1906, 1436; Gabriel N. Rosenberg, "Where Are Animals in the History of Sexuality?," *Notches*, September 2, 2014, http://notchesblog.com/2014/09/02/where-are-animals-in-the-history-of-sexuality/.
16. Harris, *Harris on the Pig*, 20.
17. William Jack, "Manufacturing Hogs," in *Special Report of the Indiana State Board of Agriculture on the Hog*, 89–91.
18. "Hogs That Make Good," *Wallaces' Farmer and Iowa Homestead*, December 14, 1917, 1. Historian Neil Oatsvall also commented on the rise of mechanical language in hog husbandry, but I locate the advent of that discourse in the nineteenth century. "Making Bacon: Death, Mechanization, and the Ecology of Pig Breeding in the United States, 1900–1960," *Essays in History* 46 (2012), http://www.essaysinhistory.com/making-bacon-death-mechanization-and-the-ecology-of-pig-breeding-in-the-united-states-1900-1960/.
19. L. H. Cooch, *500 Questions Answered about Swine* (St. Paul, MN: Webb, 1907), 23.
20. Elmer Baldwin, "Breeding of Swine—No. V," *Prairie Farmer*, February 12, 1870, 1.
21. "Rearing Hogs in Oregon," *Prairie Farmer*, April 26, 1873, 5.
22. F. D. Coburn, *Swine Husbandry* (New York: Orange Judd, 1900), 108–109.
23. Coburn, *Swine Husbandry*, 112; "Feeding Hogs on Grass," *Wallaces' Farmer and Iowa Homestead*, June 1, 1900, 15; "Fattening Hogs in the Field," *American Farmer*, July 1870, 104.
24. Tracy, "Hog Raising in the South," 19–22; "Feeding Animals—No. X, the Pig," *National Livestock Journal*, October 1877, 447.

NOTES TO PAGES 188–190

25. John M. Evvard and W. H. Pew, "Successful Swine Rations for the Corn Belt," Circular No. 26 (Ames: Agricultural Experiment Station, Iowa State College of Agriculture and Mechanic Arts, Animal Husbandry Section, March 1916), 12.
26. "The Hog as a Labor Saver," *Wallaces' Farmer and Iowa Homestead*, August 9, 1918.
27. Arthur T. Thompson, "Let Pigs Husk Some of Your Corn," *Wallaces' Farmer and Iowa Homestead*, September 19, 1942.
28. Harris, *Harris on the Pig*, 132.
29. Coburn, *Swine Husbandry*, 193, chap. 20.
30. A. W. Oliver and E. L. Potter, "Fattening Pigs for Market," Station Bulletin 269 (Corvallis: Agricultural Experiment Station, Oregon State Agricultural College, November 1930), 9.
31. Charles Dawson, *Success with Hogs* (Chicago: Forbes, 1919), 175.
32. Evvard and Pew, "Successful Swine Rations," 3–5.
33. Cooch, *500 Questions Answered about Swine*, 12.
34. "Cooking Soybeans for Hogs," *Ohio Farmer*, December 16, 1922, 22; quote from Arthur T. Thompson, "Hogmen Already Have Know-How," *Wallaces' Farmer and Iowa Homestead*, March 7, 1942.
35. "Why Hogs Need More Than Corn," *Wallaces' Farmer and Iowa Homestead*, September 5, 1942, 8.
36. Shepard, *Hog in America*, 33–37; Biggle, *Biggle Swine Book*, 47–54; Harris, *Harris on the Pig*, chapter 16.
37. *How to Make 45% More on Your Hogs!* (Sioux City, IA: Phillip Bernard Company, 1919), 4.
38. *Farm Buildings: A Compilation of Plans for General Farm Barns, Cattle Barns, Horse Barns, Sheep Folds, Swine Pens, Poultry Houses, Silos, Feeding Racks, Etc.* (Chicago: Sanders, 1905), 87.
39. "Pigs Like a Dry Bedroom," *Wallaces' Farmer and Iowa Homestead*, September 5, 1942, 5.
40. *Farm Buildings*, 88–89; William Dietrich, "The Location, Construction, and Operation of Hog Houses," Bulletin No. 109 (Urbana: University of Illinois Agricultural Experiment Station, June, 1906); T. A. H. Miller, Wallace Ashby, and J. H. Zeller, "Hog-Housing Requirements," USDA Circular No. 701 (Washington, DC: Government Printing Office, May 1944), 10–11.
41. *How to Make 45% More on Your Hogs!*, 7.
42. Chris Mayda, "Pig Pens, Hog Houses, and Manure Pits: A Century of Change in Hog Production," *Material Culture*, Spring 2004, 19–24.

43. J. D. McVean and R. E. Hutton, "Movable Hog Houses," USDA Circular No. 102 (Washington, DC: Government Printing Office, February, 1918).
44. George M. Rommell, "Pig Management," in *Uncle Sam's Farm Book*, 2nd ed., compiled by F. D. Coburn (St. Joseph, MO: News Corporation, 1912), 135.
45. "The Hog and Cleanliness," *Prairie Farmer*, October 3, 1868, 1.
46. Biggle, *Biggle Swine Book*, 119.
47. Biggle, *Biggle Swine Book*, 126.
48. A. J. Lovejoy, *Hogs: A Practical Book for the Pure Bred Swine Breeder and Farmer* (Chicago: Frost, 1919), 52.
49. Brian W. Beltman, *Dutch Farmer in the Missouri Valley: The Life and Letters of Ulbe Uringa, 1866–1950* (Urbana: University of Illinois Press, 1996), 127.
50. For a small sample of the advice about pasture, see H. Benton, "Hog Raising in the South," *Southern Cultivator*, May 15, 1906, 27.
51. Robert H. Ferrell, *Harry S. Truman: His Life on the Family Farms* (Worland, WY: High Plains, 1991), 54.
52. E. Z. Russell and J. H. Zeller, "Self-Feeding Versus Hand-Feeding Sows and Litters," US Department of Agriculture Farmers' Bulletin No. 1504 (Washington, DC: Government Printing Office, 1937), 3–6; Hayne, *Hogs for Pork and Profit*, 28–29.
53. "Cement Floor," *Wallaces' Farmer and Iowa Homestead*, March 16, 1900, 20.
54. "Feeding Floors for Fattening Hogs," *Wallaces' Farmer and Iowa Homestead*, January 26, 1917; Miller, Ashby, and Zeller, "Hog Housing Requirements," 12.
55. A. C. Spivey, "Hog Health," *Berkshire World and Corn Belt Stockman*, April 1, 1917, 11.
56. Oscar Steanson and R. H. Wilcox, "Cost of Producing Hogs in Iowa and Illinois, Years, 1921–1922," US Department of Agriculture Department Bulletin No. 1381 (Washington, DC: Government Printing Office, March 1926), 14–16.
57. E. T. Robbins, "Cheaper and More Profitable Pork Thru Swine Sanitation: A Review of the McLean County System of Swine Sanitation on Illinois Farms During 1925," Circular No. 306 (Urbana: University of Illinois Agricultural College and Experiment Station, February 1926), 2.
58. C. Clyde Jones, The Burlington Railroad's Swine Sanitation Train of 1929," *Iowa Journal of History* 57(1) (January 1959): 23–33; "Pig Crop Special" Flyers, Iowa State University Library, Special Collections Department, Paul C. Taff Papers, RS 16/3/56, box 14, folders 16, 20.
59. W. F. Ward, "Boys' Pig Clubs, with Special Reference to Their Organization in

the South," US Department of Agriculture Farmers' Bulletin No. 566 (Washington, DC: Government Printing Office, 1913), 2, 5.
60. Robert S. Curtis, "The History of Livestock in North Carolina," Bulletin No. 401 (Greensboro: North Carolina State University, 1956), 92.
61. Michael D. Thompson, "This Little Piggy Went to Market: The Commercialization of Hog Production in Eastern North Carolina from William Shay to Wendell Murphy," *Agricultural History* 74(2) (Spring 2000): 572–574.
62. H. A. Wallace, "Some Farming Tips from Iowa County," *Wallaces' Farmer and Iowa Homestead*, October 8, 1926, 6, 21.
63. "Iowa Hog Producers and the National Defense Program," Pamphlet 7 (Revised) (Ames: Agricultural Extension Service, Iowa State College, October, 1941), Iowa State University Library, Special Collections Department, Parks Library.
64. R. H. Wilcox, W. E. Carroll, and T. G. Hornung, "Some Important Factors Affecting Costs in Hog Production," Bulletin No. 390 (Urbana: University of Illinois Agricultural Experiment Station, June, 1933), 6–8. Also see H. C. M. Case and Robert C. Ross, "The Place of Hog Production in Corn-Belt Farming," Bulletin No. 301 (Urbana: University of Illinois Agricultural Experiment Station, December, 1927).
65. "They Raise Pigs Like Broilers," *Farm Journal* 80 (July 1956): 44–46.
66. Mark R. Finlay, "Hogs, Antibiotics, and the Industrial Environments of Postwar Agriculture," in *Industrializing Organisms: Introducing Evolutionary History*, ed. Susan R. Schrepfer and Philip Scranton (New York: Routledge, 2004), 242–243; J. L. Anderson, *Industrializing the Corn Belt: Agriculture, Technology, and Environment, 1945–1972* (DeKalb: Northern Illinois University Press, 2009), 92–93.
67. Dean Wolf, "What Farmers Really Think of Early Weaning," *Farm Journal*, January 1956, 42, 138; Anderson, *Industrializing the Corn Belt*, 94–95.
68. Damon V. Catron, "Swine Nutrition Research-Recent, Present, and Future" (paper, Agricultural Research Institute at the National Academy of Sciences, National Research Council, Washington, DC, October 14–15, 1957), Iowa State University Library, Special Collections Department, Damon Von Catron Papers, RS 9/11/55, box 22, folder 32; Damon V. Catron, "Recent Developments in Swine Production" (paper, Blue River Feeds Company Feeders Meeting, Indianapolis, March 24, 1959), Damon Von Catron Papers, box 22, folder 15.
69. Iowa State College, "Life Cycle Housing for Hogs" (1957), 8-8542, Iowa State

University Library, Special Collections; Damon Catron, "A Coming Way to Raise Hogs," *Successful Farming*, September 1955.

70. "Do You Grow Hogs in Confinement?," *Wallaces Farmer*, October 8, 1966,
71. Neil Oatsvall argued that by 1940 pig experts abandoned the idea of pasture for hogs "or at least that they should be fed exclusively by forage." Many experts, however, continued to counsel pasture systems into the 1960s and had advocated for grain and other feeds as a supplement for forage in the early twentieth century. See Oatsvall, "Making Bacon."
71. "Agriculture: Phrenological Pickers and Such," *Time*, October 2, 1964.
72. John Harvey, "3 Buildings, 1,200 Hogs—All Under Roof," *Successful Farming*, April 1965, 54.
73. Harvey, "3 Buildings," 54.
74. J. W. Bailey, "Don't Overlook Hogging-Down Corn," *Successful Farming*, September 1963, 74.
75. A. H. Jensen, B. G. Harmon, G. R. Carlisle, and A. J. Muehling, "Management and Housing for Confinement Swine Production," Circular No. 1064 (Urbana: University of Illinois College of Agriculture, Cooperative Extension Service, November 1972), 3.
76. A. C. Dale and J. E. Mentzer, "Swine Waste Management and Disposal," Extension Service Paper No. 51 (West Lafayette, IN: Purdue University Extension, September, 1969), 1.
77. Karen McMahon, "Premium Standard Farms," *National Hog Farmer*, May 1, 1988, http://nationalhogfarmer.com/mag/farming_premium_standard_farms.
78. Don Jones and Alan Sutton, "Manure Storage Systems, ID-352" (West Lafayette, IN: Purdue Extension, August 2007).
79. Nigel Key, William D. McBride, Marc Ribaudo, and Stacy Sneeringer, "Trends and Developments in Hog Manure Management: 1998–2009," Economic Information Bulletin No. 81 (Washington, DC: USDA Economic Research Service, September 2011), 11.
80. D. L. "Hank" Harris, *Multi-Site Pig Production* (Ames: Iowa State University Press, 2000), 3–13.
81. Kirk Clark, Chris Hurt, Ken Foster, and Jeffrey Hale, "All-In/All-Out Production," in *Positioning Your Pork Operation for the 21st Century* (West Lafayette: Purdue University Cooperative Extension Service, n.d.).
82. "Artificial Breeding Field Tests Started," *Wallaces Farmer*, March 19, 1960; "What Progress in Artificial Breeding for Swine?," *Wallaces Farmer*, June 16,

1962; Chester Peterson, Jr., "What's Happening with Swine AI?," *Successful Farming*, July 1962, 31, 44.

83. Betsy Freese, "AI Gets Another Try," *Successful Farming* (Mid-March 1994): 28.

84. Mark J. Estienne and Allen F. Harper, "Using Artificial Insemination in Swine Production: Detecting and Synchronizing Estrus and Using Proper Insemination Technique," 414-038 (Blacksburg: Virginia Tech University, May 1, 2009); Nigel Key and William McBride, "The Changing Economics of U.S. Hog Production" (USDA-ERS Economic Research Report 52; Washington, DC: USDA and ERS, December, 2007), 12.

85. Zachary Henson and Conner Bailey, "CAFOs, Culture and Conflict on Sand Mountain: Framing Rights and Responsibilities in Appalachian Alabama," *Southern Rural Sociology* 24(1) (2009): 153–174.

86. Ralph Watkins, "Odor Complaints Force Custom Feedlot Shutdown," *National Hog Farmer*, April 1978, 24, 26.

87. Betsy Freese, "Talking Back on Hogs," *Successful Farming*, August 2003, 60.

88. Carolyn Johnsen, *Raising a Stink: The Struggle over Factory Hog Farms in Nebraska* (Lincoln: University of Nebraska Press, 2003), 1–3.

89. Joe Holley and Evan Smith, "Hog-tied," *Texas Monthly* 25(12) (December 1997): 26.

90. Susan S. Schiffman, Elizabeth A. Sattely-Miller, Mark S. Suggs, and Brevick G. Graham, "Mood Changes Experienced by Persons Living Near Commercial Swine Operations," in *Pigs, Profits, and Rural Communities*, ed. Kendall M. Thu and E. Paul Durrenberger (Albany: State University of New York Press, 1998), 87–88; Steve Wing, Rachel Avery Horton, Stephen W. Marshall, Kendall Thu, Mansoureh Tajik, Leah Schinasi, and Susan S. Schiffman, "Air Pollution and Odor in Communities Near Industrial Swine Operations," *Environmental Health Perspectives* 116(10) (October 2008): 1366–1367; Rachel Avery Horton, Steve Wing, Stephen W. Marshall, and Kimberly A. Brownley, "Malodor as a Trigger of Stress and Negative Mood in Neighbors of Industrial Hog Operations," *American Journal of Public Health* 99(S3) (2009): S610–S615.

91. Christine Schrum, "Hog Confinement Health Risks," *Iowa Source*, August 5, 2005, http://www.iowasource.com/health/CAFO_airqu_0805.html.

92. Dana Cole, Lori Todd, and Steve Wing, "Concentrated Swine Feeding Operations and Public Health: A Review of Occupational and Community Health Effects," *Environmental Health Perspectives* 108(8) (August 2000): 685.

93. Vukina Tomislav, Fritz Roka, and Raymond B. Palmquist, "Is It Costly to Be a Hog Farm Neighbor?," *Choices: The Magazine of Food, Farm and Resource Issues*

11(1) (1996); J. H. Conrad and V. B. Mayrose, "Animal Waste Handling and Disposal in Confinement Production of Swine," *Journal of Animal Science* 32(4) (1971): 811–815.

94. Joseph A. Herriges, Silvia Secchi, and Bruch A. Babcock, "Living with Hogs in Iowa: The Impact of Livestock Facilities on Rural Residential Property Values," *Land Economics* 81(4) (November 2005): 530–545.

95. Jennifer Dukes Lee, "Hog Odor Battles Head to Court," *Des Moines Register*, March 21, 2004, 1A, 4A.

96. "The Brown Lagoon," *Economist* 336 (September 2, 1995), 24; Michael Satchell, "Hog Heaven—and Hell," *U.S. News & World Report* 120(3) (January 22, 1996).

97. Amy R. Sapkota, Frank C. Curriero, Kristen E. Gibson, and Kellogg J. Schwab, "Antibiotic-Resistant Enterococci and Fecal Indicators in Surface Water and Groundwater Impacted by a Concentrated Swine Feeding Operation," *Environmental Health Perspectives* 115(7) (July 2007): 1040–1045.

98. Michael T. Meyer, J. E. Bumgarner, J. V. Daughtridge, Dana Kolpin, E. M. Thurman, and K. A. Hostetler, "Occurrence of Antibiotics in Liquid Waste at Confined Animal Feeding Operations and in Surface and Ground Water," in *Effects of Animal Feeding Operations on Water Resources and the Environment, Proceedings of the Technical Meeting U.S. Geological Survey Open-File Report 00-204* (Reston, VA: US Geological Survey, 2000). For more about CAFOs and the Ogallala Aquifer, see John Opie, *Ogallala: Water for a Dry Land*, 2nd ed. (Lincoln: University of Nebraska Press, 2004), 160–163.

99. Johnsen, *Raising a Stink*, 31–33.

100. Thompson, "This Little Piggy Went to Market," 582–583; Paul B. Stretesky, Janies E. Johnston, and Jeremy Arney, "Environmental Inequity: An Analysis of Large-Scale Hog Operations in 17 States, 1982–1997," *Rural Sociology* 68(2) (2003): 233; Maria C. Mirabelli, Steve Wing, Stephen W. Marshall, and Timothy C. Wilcosky, "Race, Poverty, and Potential Exposure of Middle-School Students to Air Emissions from Confined Swine Feeding Operations," *Environmental Health Perspectives* 114(4) (April 2006): 591–596.

101. Charles Mahtesian, "Battling Boss Hog," *Governing* 9 (April 1996): 30–33.

102. Donald L. Bartlett and James B. Steele, "The Empire of the Pigs," *Time*, June 24, 2001; Heather Williams, "Fighting Corporate Swine," *Politics and Society* 34(3) (September 2006): 388.

103. Stretesky, Johnston, and Arney, "Environmental Inequity," 233–234.

104. Kendall Thu and E. Paul Durrenberger, "North Carolina's Hog Industry: The Rest of the Story," *Culture and Agriculture* 14(49) (March 1994): 23.
105. Mark S. Honeyman and Michael D. Duffy, "Iowa's Changing Swine Industry" (Animal Industry Report: AS 652, ASL R2158, Iowa State University, 2006).
106. "Meatpacking" (US Department of Labor, Occupational Safety and Health Administration), https://www.osha.gov/SLTC/meatpacking/index.html; Lynn Waltz, "Slaughterhouse '05," *Port Folio Weekly*, July 5, 2005, 19–20.
107. Joby Warrick and Stuart Leavenworth, "Waste Spill Revives Hog Legislation. House Looks Again at Farm Regulations," *Raleigh News and Observer*, June 29, 1995.
108. Jeff Tietz, "Boss Hog: The Rapid Rise of Industrial Swine," in *The CAFO Reader: The Tragedy of Industrial Animal Factories*, ed. Danial Imhof (Berkeley: Watershed Media, 2010), 121; "Key Industries: Hog Farming" (Learn NC, North Carolina Digital History, University of North Carolina School of Education), http://www.learnnc.org/lp/editions/nchist-recent/6257.
109. Patty Cantrell, "Is the Family Farm an Endangered Species?," *Ms.* 7(5) (March/April 1997).
110. Eric Whitney, "Can a Hog Farm Bring Home the Bacon?," *High Country Times*, November 8, 1998.
111. Paul Tolme, "They Fought for Their Tribe—and Won," *National Wildlife* 42(1) (December 1, 2003): 16–17.
112. Cited in Johnsen, *Raising a Stink*, 18.
113. Kelly Donham and Kendall Thu, *Understanding the Impacts of Large-Scale Swine Production: Proceedings from an Interdisciplinary Scientific Workshop* (Des Moines, IA, June 29–30, 1995), 1–6.
114. Donham and Thu, in *Understanding the Impacts*, 12, 14, 17.
115. Peter S. Thorne, in *Understanding the Impacts*, 153–158, 167.
116. Paul Lasley, in *Understanding the Impacts*, 118, 127–128, 135–137.
117. Lorah Slaton, "Lawsuits over Odors from Missouri Hog Confinements Are Settled," *Kansas City Star*, August 30, 2012; Kelsey Volkmann, "Hog Farm Operators Ordered to Pay $11 M for Odor," *St. Louis Business Journal*, March 5, 2010, http://www.bizjournals.com/stlouis/stories/2010/03/01/daily74.html?ed=2010-03-05.
118. Bridget Huber, "Law and Odor: How to Take Down a Terrible-Smelling Hog Farm," *Mother Jones*, May–June 2014, http://www.motherjones.com/environment/2014/04/terrible-smell-hog-farms-lawsuits.

119. Melinda Wenner Moyer, "How Drug-Resistant Bacteria Travel from the Farm to Your Table," *Scientific American*, December 2016, https://www.scientificamerican.com/article/how-drug-resistant-bacteria-travel-from-the-farm-to-your-table/.

## CODA

1. Joe Vansickle, "Unveiling the Carbon Footprint Model," *National Hog Farmer*, May 15, 2011, 47.
2. Quoted in Nicolette Hahn Niman, *Righteous Porkchop: Finding a Life and Good Food Beyond Factory Farms* (New York: Harper, 2009), 241.
3. Wanda Patsche, "What I Wish People Knew about Pig Farming," *Minnesota Farm Living*, October 6, 2014, http://www.mnfarmliving.com/2014/10/wish-people-knew-pig-farming.html.
4. David K. C. Cooper, "A Brief History of Cross-Species Organ Transplantation," *Baylor University Medical Center Proceedings* 25(1) (2012): 50.
5. Sharon Begley, "Charlotte Said It Best: Some Pig," *Newsweek* 122(17) (October 25, 1993).
6. Lawrence M. Fisher, "Down on the Farm, a Donor: Breeding Pigs That Can Provide Organs for Humans," *New York Times*, January 5, 1996, D1.
7. Cooper, "Brief History," 53–54.
8. Sheryl Gay Stolbert, "Could This Pig Save Your Life?," *New York Times*, October 3, 1999, SM46.
9. Jonathan S. Allan, "From Pigs to People," *New York Times*, October 2, 2001, F5.
10. Sara Reardon, "New Life for Pig-to-Human Transplants," *Nature*, November 10, 2015, http://www.nature.com/news/new-life-for-pig-to-human-transplants-1.18768.
11. Antonio Regalado, "Surgeons Smash Records with Pig-to-Primate Organ Transplants," *MIT Technology Review*, August 12, 2015, http://www.technologyreview.com/news/540076/surgeons-smash-records-with-pig-to-primate-organ-transplants/.
12. Nicholas Wade, "New Prospects for Growing Human Replacement Organs in Animals," *New York Times*, January 1, 2017, https://www.nytimes.com/2017/01/26/science/chimera-stemcells-organs.html; Gina Kolata, "Gene Editing Spurs Hope for Transplanting Pig Organs into Humans," *New York Times*, August 10, 2017, https://www.nytimes.com/2017/08/10/health/gene-editing-pigs-organ-transplants.html?_r=0.

# INDEX

*Page numbers in italic refer to figures.*

African Americans, 21, 74, 81, 153, 157, 160–62, *165*, 173, 175–76, 183
AIAO (all-in/all-out) practice, 205, 214
air quality, 207, 210, 213–14, 218, 219
Allan, Jonathan, 222
alligator pigs, 41, 54
America
   Euro-American taste, 64–71
   feral hogs in rural landscape, 30–34
   first swine in North America, 7–8
   future of hogs, 217–23
   gehography, 11–34
   history of hogs, 1–10, 11–12, 217–18
   native encounters with swine, 12–16
   pigs, pork, and Civil War, 18–20
   porcineograph, 11, *13*
   porkopolis to Pig War, 17–18
   post-emancipation, 21–23
   postwar pigs, 23–30
*American Agriculturalist*, 17, 53
*American Husbandry*, 38, 106
*American Swine Breeder, The* (Ellsworth), 55, 137
Anderson, David, 79
Anderson, Virginia DeJohn, 8, 11, 16
antibiotics, 166, 172, 199, 208, 210, 214–15, 219
*Arator* (Taylor), 53
Armour, 128, 129, 164
army, 19, 20, 78–82, 95, 124, 132
Arthur, Chester, 127
artificial insemination (AI), 205–6
Asiatic cholera, 135–36

Baby Fae, 221
back fat, 168, 169, 172
bacon
   definitions, 69, 119
   fattening hogs, 45, 188, 189
   industrial America, 156, 157, 159
   markets, 105, 107, 119–21, 124, 128, 129
   persistence and threats, 160, 162, 163
   post–Civil War rural America, 153, 154
   United States of bacon, 177, 178–80
   wartime pork, 20, 78, 79, 81
   working people's food, 61–63, 67–71, 72–74
Baconfest, 179
BAI. *See* Bureau of Animal Industry
Baker, W. E., 11
Bakewell, Robert, 59
Baltimore, Lord, 37
Bangor, Maine, 91, 92
barbeque, 64–65, 75, 161, 166
barreled pork, 111, 114, 155, 159
Bartlett, H., 91
Batali, Armandino, 172
Bedford, Duke of, 55
beef
   bacon and white meat, 152, 166, 167, 173, 174
   industrial America, 155, 156, 158, 159
   markets, 107, 130, 132
   pork persistence and threats, 160, 162
   post–Civil War rural America, 153–55
   working people's food, 62, 63, 66–69, 72–75, 77–78, 82
Behr, Edward, 171
Bell Farms, 211–12
Bergland, Bob, 149
Berkshire breed, 54, 59, *60*, 123, 172, 185
*Berkshire World and Corn Belt Stockman*, 144, 146

Bernard, Mabel, 207
BIA (Bureau of Indian Affairs), 211, 212
Biggle, Jacob, 184, 192
Billings, Frank S., 141, 142
Bird, Isabella, 113
Bispham, Joseph, 43
Black Hawk, *60*
Black Power movement, 175
Blake, Carl, 172
Blue Ribbon Bacon Festival, 179
boar, wild, 9, 31, 32, 36, 58–59
Bodenchuk, Michael, 32
Bogue, Allan, 22
Bosque, Fernando del, 8
Bost, Theodore, 76
Boston, 69, 91, 92, 143
Bowen, Bill, 206
breeding hogs, 52, 54–60, 184–86, 197, 205–6
brine, 62, 106, 111, 114, 119, 155, 159
Brock, Sean, 172
Brooklyn, 88, 91, 92
Bureau of Animal Industry (BAI), 136, 141–47, 191, 195
Bureau of Indian Affairs (BIA), 211, 212
burgers, 166, 178, 180
Burlend, Rebecca, 43
Burnaby, Rev. Andrew, 67, 107
Burton, Robert, 63
Burwell, Carter, 107
Burwell, Nathaniel, 42
butchering, 70, 114, 155, 157–58, 162
Butler, Earl, 196–97
Butz, Bob, 31
Byrd, William, 39, 66–67, 71

CAFOs. *See* concentrated animal feeding operations
California, 11, 94, 98
Calyo, Nicolino Vicomte, 90
Caribbean, 64, 69, 106, 134
Carpenter, Helen, 72–73

Carr, Isaac, 155
Carter, Robert, 65
Cartmell, Robert, 135, 139
Catron, Damon V., 199, 202
cattle, 14–16, 64, 66, 69, 77, 136, 141, 154, 218
charcuterie, 152, 218
checkoff, 133
Cherokee people, 15–16
Chesapeake settlers, 9, 14, 38, 39, 42, 43, 65, 68
Chicago, 115, 118, 126
chicken, 70, 152, 162, 166–67, 171–73, 180
China, 7, 133–34
Chinese hogs, 59, 60, 123, 172
chitterlings/chitlins, 74, 160, 161, *164, 165,* 175, 176
cholera, 88, 135–36
See also hog cholera
Christianity, 15, 16, 63
Cincinnati ("Porkopolis"), 17, 110, 112–18, 120, 123, 126, 135, 137
civil rights movement, 175
Civil War, 12, 18–20, 34–36, 60, 62, 78–82, 124, 136, 153, 184
Clay, Cassius, 44
Clayton, Rev. John, 67
cleanliness, 6, 58, 191, 192, 194
clear pork, 119, 120
Clemens, Paul, 107
closing the range, 52, 53, 54, 60, 154, 183
Coburn, Foster, 187, 188
Colden, Mayor Cadwallader, 87
Cole, J. W., 171
Cole, Robert, 38–39, 43
colonialism, 7–8, 10, 12, 14–16, 37–38, 47, 52, 62–65, 67–69, 71, 106–7
Combs, Bob, 103–4
"Coming Way to Raise Hogs, A" (Catron), 199, 200
concentrated animal feeding operations (CAFOs), 181–82,

206, 207, 209, 210–12, 214, 219
Confederacy, 12, 18–21, 78, 79, 81, 152, 153
confinement, 34, 181, 199, 200–204, 206–16
Connaway, John, 145
Connecticut, 101–3, 107
Continental Grain, 211, 214
contracting, 25–28
controlling hogs, 3–5, 31–33, 46–50
Cook, Captain James, 7, 106
Cooper, David, 222
corn
  feeding hogs, 12, 42–45, 56–57, 128, 188–89
  history of hogs, 17, 22, 23
  markets, 111, 118, *125,* 126, 128, 130
Corn Belt, 24, 27, 105, 131
Cornish, John Hamilton, 75
cotton, 73, 78, 111, 121, 153, 158, 183, 196
Covina, California, 94
cracklings, 123
Creighton, James, 76
Cronon, William, 115
crossbreeding, 186, 197
Cuba, 7
culatelli (Italian-style ham), 172
Cunningham, John, 21
Cutlar, Lyman, 17–18

dairy waste, 189
Danhof, Clarence, 53
Danish bacon, 128, 129
Dean, Mary (Pig Foot Mary), 161
Deer Island, 143
defense ham, 132
Delaware, 37, 52, 70, 157
Demuth, Albert, 82
de Schweinitz, Alexander, 142, 143
de Soto, Hernando, 7, 16, 63

# INDEX

Detroit, *165*
Dickens, Charles, 85–86
diet (human)
  American preferences, 152–53
  chicken and white meat, 165–72
  civil rights to soul food, 174–77
  Euro-American taste, 64–71
  European taste, 62–64
  future of hogs, 217, 218
  history of hogs, 5, 7
  pork and lard in industrial America, 155–59
  pork and post–Civil War rural America, 153–55
  pork persistence and threats, 159–65
  postwar export markets, 133
  rehabilitating pork and lard, 172–74
  United States of bacon, 177–80
Dietrich, William, 186
disease
  cholera, 88, 135–36
  epidemics, 136, 137, 218
  new diseases, 150–51
  science and swine, 192, 194–95, 205, 207, 214
  trichina, 98–99, 127–28, 162–63, 170
  urban pigs, 85, 88, 93, 98–99
  *See also* hog cholera
Ditchkoff, Stephen, 30
Dix, General John, 81
Dodge, Grenville, 20
Dodge, John, 107
Dorset, Marion, 136, 142, 143, 144, 146
Dreiser, Theodore, 156
drovers, 108–10
Duden, Gottfried, 44
Duffy, John, 91
Duffy, Mike, 220
Duroc-Jersey breed, 185, 191, 197

Dutch settlers, 8, 9, 42, 69, 91–92

Earle, Rebecca, 63–64
earthen manure structures, 203–4, 208, 209
*E. coli*, 210, 219
Edelson, Max, 38
Eggleston, Benjamin, 115
Ellsworth, Henry William, 45, 55, 56, 58, 137
Emergency Hog Marketing Program, 130–31
enclosures, 42, 47–49, 51–55, 57, *138*, 192, 202
English hogs, 128, 129
English market, 120–21, 124, 132
English settlers
  English diet, 63, 65–67, 69, 77, 82, 152, 188
  first swine in North America, 8
  hog husbandry, 37, 38, 54, 55
  native encounters with swine, 12, 13, 15
  pork in Atlantic marketplace, 106, 107
  working people's food, 62, 65, 66, 69, 82
enslaved people, 71–74, 81, 107, 121–22, 153
environmental impact, 212, 214, 218, 220
Environmental Protection Agency, 103, 208, 218
Eringa, Ulbe, 143, 193
Essig, Mark, 5
European hogs, 58–59
European market, 127, 128, 130, 133
European settlers, 3, 7–11, 17, 37, 54–55, 62–64, 82, 162
Evelyn, Robert, 8
Evvard, John, 95, 188

fall fattening, 42–46, 56
*Farm Journal*, 28, 198

farrowing, 8, 193, 195, 199–202
fast food, 166, 174, 178–79, 180
fat, 122, 132, 153, 157, 163, 166–69, 171–74, 178
fattening hogs, 17, 42–46, 54, 56, 57, 188
feeding hogs
  corn feeding, 12, 42–45, 56–57, 128, 188–89
  flooring, 193, 194, 199
  garbage feeding, 77, 93–104, 148, 162–63
  improved husbandry, 54–57, 60
  pasture feeding, 42, 43, 45, 49, 52–53, 128, 172, 183, 186–87, 193
  self-feeding, 188, 193
fencing, 14, 33, 35–36, 45, 47–48, 50–54, 183
feral hogs, 3, 7, 12, 30–34, 36, 40, 218
Flagg, Gershom, 40
Flatbush, 91–92
Florida, 7, 11, 30, 139
Fontana Farms, 95, 96, *96*, *97*, 100
forests, 37–38, 40–42, 50–51, 53, 63
Foster, Matthew, 40
France, 127, 128, 129
Freeman, Warren, 78
free-range hogs, 37–38, 46, 48–50, 54, 59–60, 85–87, 104, 182–83
Freese, Betsy, 29
French settlers, 107, 108
Fuller, Thomas, 37–38

Gal molecule, 221–22, 223
Gamble, James, 122
game, 16, 31, 65, 66, 67, 68, 74
garbage feeding, 77, 93–104, 148, 162–63
Garrard, George, 59
Gates, Paul, 19
Gebby, Margaret, 139, 155, 158

279

genetics, 205, 219, 222, 223
Georgia, 10, 21, 33, 47, 74, 160, 182
German settlers, 8, 23, 69, 70, 156
Germany, 127, 128, 129, 130
germ theory, 141, 194, 195
gestation, 6, 34, 172, *201*, 202
Gignilliat, John, 127
Gray, Lewis, 109
Grayson, William J., 79
Great Depression, 131, 147, 152, 164
Great Dismal Swamp, 41
Great Hog Swindle, 124
Great Migration, 161, 176
Great Plains, 28, 34
greenfields, 12, 28, 34
greenhouse gas emissions, 218
Griffin, Charles, 17–18
groundwater contamination, 103, 208, 209, 219
Guymon, Oklahoma, 209, 210

Hahn, Steven, 183
ham
  making bacon and white meat, 154–55, 157, 159, 163–64, 168–69, 176–77, 180
  markets, 107, *113*, 114, 119, 120, 124, 129
  porcineograph, 11
  working people's food, 61, 67–69, 71, 72, 75, 76
Hamer, Fannie Lou, 175
Hamilton, Alexander, 70
Hamilton, Robert, 201, 202
Hardee's, 178
Hardin County, Tennessee, 25, 36, 186
*Harper's Weekly*, 72, 77, *111*
Harriet, Christian, 87, 88
Harris, Jessica, 176
Harris, Joseph, 184, 186, 188
Harrison, Benjamin, 128
Hartford, Connecticut, 101, 102–3

*Hartford Courant*, 102–3
Hawaii, 7, 31, 106, 177
Haws, Bill, 27
Hazel, Lanoy, 169
headcheese, 122, 156, 157, 175
health
  hogs, 194–98
  human diet, 166, 171, 174, 180
  public health, 91, 92, 94, 127, 128
  workers, 210, 213, 219
  xenotransplantation, 220–23
heart disease, 166, 221–23
Hempstead, Stephen, 44
heritage pork, 172
Heston, Edward, 56
Hicks, Sarah Frances, 75
Hilliard, Sam Bowers, 121
Hillman, Jimmye, 160
Hillstown Road piggery, 101–2
Hispanic population, 157, 173
Hispaniola, 7
Hogberg, Maynard, 171
hog cholera
  Civil War, 78
  control, 144–49
  history of hogs in America, 20, 26, 218
  origins and cures, 136–40
  science and swine, 192, 197
  swine plagues, 135, 136, 150, 151
  urban pigs, 95, 97–99, 103
  USDA, 140–44
Hog Cholera Research Station, 144, *146*, *147*
hog dogs, 182–83
hog drives, 91, 108–10, *111*
hog farming
  future of hogs, 220
  history of hogs, 2, 3, 5, 9, 12, 34
  large-scale, 28, 34, 150–51, 209–14, 218, 220
  post-emancipation, 22–23
  postwar pigs, 23–26, 28, 29

working people's food, 63, 72
  *See also* hog husbandry
hogging down, 44–45, *46*, 188, 202
hog houses, 57, 189–91, *194*, 197
hog husbandry, 35–60
  closing the range, 50–54
  feeding, breeding, shelter, 54–60
  historical context, 5, 6, 15
  open range, 37–41
  overview, 35–37
  rise of fall fattening, 42–46
  science and swine, 182, 184, 186, 191, 196, 197
  yokes, rings, and enclosure, 46–50
Hog Island, 14, 15, 47
hog number 844, 143
hogs
  American gehography, 11–34
  future in America, 217–23
  history in America, 1–10
  home on the range, 35–60
  making bacon and white meat, 152–80
  markets and marketing, 105–34
  science and swine, 181–216
  swine plagues, 135–51
  urban pigs, 83–104
  working people's food, 61–82
*Hogzilla* documentary, 33
Holland Land Purchase, 40
Holley, Mary Austin, 41
Hooker, Richard, 155
Hoover, Herbert, 129, 130
horizontal wheel, 115
Hormel, 132, 167, 177
Hormel, Jay, 164
Horowitz, Roger, 159
housing hogs, 57, 189–91, *194*, 197
Howard, Paine, 44
Howells, William Cooper, 43, 76
Hubbard, Bolling, 156
Hudson, John, 126
hunting hogs, 14, 31–33

280

INDEX

Hurt, Doug, 118
husbandry. *See* hog husbandry

Illinois, 41, 107, 109, 139, 162
illumination, 106, 122, 123, 124
immigration, 23, 162, 163, 219–20
immunization, 95, 97, 143, 145, 147
Indiana, 24, 108, 112, 120, 135, 139
indigenous people, 3, 12–16
industrialization, 24, 155–59, 220
infections, 28, 193, 195, 205, 222
inoculation, 141, 148
inspection law, 127, 128
insulin, 223
Iowa, 17, 23–25, 98, 133, 139, 142, 144, 146, 200, 214
Iowa Porkettes, 170
Iowa Pork Producers, 29, 170
Irish market, 126–27
Irish settlers, 88, 92, 163
Islam, 64
isowean (isolated weaning), 205
Italy, 127, 129
Iyotte, Eva, 212

Jamaica, 64, 107
James Towne (later Jamestown), 8, 14, 38, 47, 66
Janes, Thomas P., 21, 182
Japan, 29, 133
Jeckel, Russ, *201*, *203*
Jefferson, Thomas, 71
Johnson, Nathanael, 171
Johnson, Oliver, 108–9
Jones, George W., 112–13
Jones, Robert L., 51, 137
Josselyn, John, 106
Judas hogs, 32
Jukes, Thomas Hughes, 199

Kalm, Pehr, 42
Kansas City, 161
Kennedy, John F., 148

Kent Island, Maryland, 47
Kentucky, 12, 18, 19, 78, 111, 112, 121, 124–25, 137
Khruschchev, Nikita, 132
Kieft, William, 15
Kilbourne, F. L., 142
Kimball family, 155, 157
King, Edward, 154
King Philip's War (1675), 15, 16
Kirst, Christian, 156
Knight, Sarah Kemble, 70
Koren, Elisabeth, 61

Lakier, Alexander, 113
Landis, Leo (the Bacon Professor), 179
land pikes, 41, 45, 54, 184
Landrace hog, *170*
land sharks, 9, 41, 54
lard
  future of hogs, 217, 218
  history of hogs, 2, 5, 6
  making bacon and white meat, 156–58, 160–62, 167, 169, 173–74, 176
  markets and marketing, 105, 106, 114, 119, 120, 122–24, 128–30
  working people's food, 64, 72
lard oil, 121, 123, 124, *125*
large-scale hog farming, 28, 34, 150–51, 209–14, 218, 220
Las Vegas, 103
Latrobe, Benjamin Henry, 72, 74
Lawrence, John D., 25
leaf lard, 122
lean meat, 163, 168, 171–73, 188
Leslie, Eliza, 77
Levenstein, Harvey, 155
Lewger, John, 37
L'Hommedieu, Ezra, 56
lighting, 122, 123, *125*
Lincoln, Abraham, 19, 79
Lindgren, Ida, 154
Linton, John, 54
"Little Spot," 88

Lombardo brothers, 101–2
Long Beach, 94
Long Beach Island, 47
Los Angeles, 94, 95, *96*, *97*, 100
Louisiana, 107, 121, 139
Louisville, Kentucky, 124
Lovejoy, A. J., 191, 193
Lowry, E. F., 145
lubricants, 2, 121, 122, 123
luncheon meat, 165, 177, 180

machines, hogs as, 182, 186
*Maine Farmer*, 55, 58
Mallory, James, 21, 57, 78, 122, 139
Mangalitsa hog, 173
manure management
  hog husbandry, 42, 53, 58
  lagoons and pits, 23, 28, 203–4, 208, 211, 213
  odor, 100, 103, 219
  science and swine, 193, 202–4, 208, 209, 211, 213
  urban pigs, 84, 86, 91, 95, 100, 103
marbling, 171, 172
marked pigs, 50, 51
marketing campaigns, 105, 132–33
market refuse, 93, 94, 96
markets, 105–34
  hog drives and drovers, 108–10
  making markets, 118–26
  overview, 105–6
  pork and lard in Atlantic marketplace, 106–7
  Porkopolis, 110–18
  postwar markets, 131–34
  trade war, world war, and Great Depression, 126–31
Martineau, Harriet, 75
Maryland, 38–39, 66, 107
Marzialo, Nicholas, 102
Massachusetts, 14, 47, 49, 76, 95, 107
Massie, William, 52

mast, 38, 41, 42, 51, 63, 84
Maverick, Samuel, 106
*Mayflower* (ship), 9
McBirney, John, 142
McCoy, Renee, 176–77
McDonald's, 172
McGovern, George, 166
McKinley tariff, 128
McLean system of sanitation, 194–95, 196, 197, *198*
McNeur, Catherine, 90
meat consumption, 62–69, 74, 85, 106–7, 132, 134, 152–53, 155, 160–61
Meat Inspection Act, 128
meat-type hogs, 167–69, *169*, *170*, 205
Mecklin, Augustus, 35
Mednansky, Oleta, 212
Meeker, David, 29
Melosi, Martin, 93
Menendez de Aviles, Pedro, 7
mess pork, 119, 120, 124
Mexico, 7, 64, 133, 174
Miantonomi (Narragansett Chief), 12, 13
Michaux, Francois, 76
middle class, 77, 90, 100, 103, 155–56, 163, 176
Middleton, Richard, 214
Midwest, 22, 23, 24, 28, 173
Miller, Lewis, *48*, *59*
Miller, Sid, 33
Milwaukee, 93, 118, 123
Minnesota, 23, 126
Mississippi, 78, 108, 126, 139, 185, 188
Missouri, 76, 139, 145, 155, 156, 211, 214
Mizelle, Brett, 5
Mobley, Joe, 79
molasses, 75, 77, 86, 153, 160
Moore, A. C., 35
Moore, Varanus A., 142
mortality rate of hogs, 2, 197, 199, 205

Mount, William Sidney, 3–5
Muhammad, Elijah, 175
multisite production, 205
Murphy, Wendell, 26, 27
Murrow, Edward R., 132
mutton, 62, 63, 66–69, 72, 74, 77, 82, 152, 155

Narragansett Bay, 47
Narragansett people, 12, 15
National Farms, 27, 28
National Pork Producers Council (NPPC), 29, 171
Nation of Islam, 175
Native Americans, 7, 12–16
*Nature's Metropolis* (Cronon), 115
Nebraska, 23, 27, 28, 141, 209
New England, 9, 13–15, 45, 56, 67–68, 76–77, 85, 96–97, 106, 156
*New England Farmer*, 54, 58
New Jersey, 31, 43, 49, 70, 97
Newlands, Francis, 154
New Mexico, 8, 64, 157
New Netherland, 9, 15, 42
New Orleans, 107, 110, 120, 121, 126
"new pork," 169–70, 171
New World, 3, 7, 8, 10, 37, 63–65, 107
New York, 40, 45, 69, 77, 83–90, 118–19, 162–63
*New York Times*, 88, 222, 223
"Nickels for Profits," 133
Niles, William B., 143, 144, 146
Niman Ranch, 173, 216
Nimmo, Joseph, 127–28
Nine-Point Program, 148
N. K. Fairbank Company, *125*, 158
North American Free Trade Agreement, 133
Northbank Plantation, 51
North Carolina, 26–28, 39, 66–67, 176, 208–9, 211, 214
Northrop, Lucius B., 19

Norton, John (Teyoninhokovrawen), 16
notching, 50, 51

obesity, 163, 174
odor of hogs, 28, 86, 91–94, 100–104, 206–7, 214, 219
offal, 77, 83, 84, 89, 133, 162, 163
Oglethorpe, James, 47
Ohio, 51, 76, 86, 108–12, 118, 120, 126, 135, 137
oils, 121, 123–25, 158, 167, 172, 174
Oklahoma, 29, 209–10
"Old Seed Corner" (writer), 59–60
oleomargarine, 129
Oliver, Sandra, 65
Olmstead, Alan, 137
Olmsted, Frederick Law, 41, 44, 74–75, 114, 120
Oñate, Juan de, 7
open range, 16, 36, 37–41, 51–55, 60, 182–83
Opie, Frederick Douglass, 160
orchards, 45, 46, 53, 140
organ transplantation, 220–23
Oriental swine, 59, 60, 123, 172
Ossabaw Island hogs, 9, *9*, 172
"Other White Meat, The" (campaign), 133

packing industry, 27, 29, 72, 110–18, 126, 129, 210, 219
pannage. *See* mast
parasites, 98, 193, 195
Parkes, Samuel, 54
Parkinson, Richard, 41
Parren, N. H., 139
Pasteur, Louis, 141
pasture feeding, 42, 43, 45, 49, 52–53, 128, 172, 183, 186–87, 193
Patience Island, 47
PDCoV (porcine delta coronavirus), 151

# INDEX

PEDv (porcine epidemic diarrhea virus), 150, 151, 215
penning, 37, 39, 44, 53, 58, 183, 190, 191
Pennington, James, 73–74
Pennsylvania, 11, 42–43, 69, 70, 75–76, 107, 129, 157
Perry, Henry ("the Barbeque King"), 161
Philadelphia, 43, 46, 86, 157
Philipp, Barbara, 207
Phillips, Abraham, 43
Phinney, Elias, 55
pickled pork, 11, 62, 68, 69, 79
*Pig and the Public, The* (film), 168
pig banks, 175
Pig Clubs, 196
Pig Crop Special, 196
Pig Foot Mary (Mary Dean), 161
piggeries, 89, 91–94, 101–4
Piggery War (1859), 89
Pigneyland, 181, 182
pigs. *See* hogs
*Pigs: From Cave to Corn Belt* (Towne and Wentworth), 5
"pigs versus people" fight (1954), 102–3
pig wars, 15, 17–18, 88, 89
pileau, 74
Plager, Carroll, 167
plantations, 21, 44, 51, 73–74, 107, 121, 122, 153, 183
Poland China breed, *168*, 185
Pollard, Benjamin, 51
pollution, 28, 99, 204, 214
population of hogs, *18*, 21–23, 132, 137
porcine delta coronavirus (PDCoV), 151
porcine epidemic diarrhea virus (PEDv), 150, 151, 214
porcineograph, 11, *13*
porcine reproductive and respiratory syndrome (PRRS), 150, 151, 214

pork
  Atlantic marketplace, 106–7
  chicken and white meat, 165–72
  civil rights to soul food, 174–77
  Civil War, 78–82
  consumption per capita, 133, *167*
  Euro-American taste, 64–71
  European taste, 62–64
  future of hogs, 217–19
  history of hogs, 2–3, 5–6, 8, 12, 16–18, 21
  industrial America, 155–59
  making bacon and white meat, 152–80
  making markets, 118–26
  making republic of porkdom, 72–77
  markets and marketing, 105–34
  persistence and threats, 159–65
  pigs, pork, and Civil War, 18–20
  porcineograph, 11, *13*
  Porkopolis, 110–18
  post–Civil War rural America, 153–55
  rehabilitating pork and lard, 172–74
  trade war, world war, and Great Depression, 126–31
  United States of bacon, 177–80
  urban pigs, 96, 98, 104
  working people's food, 61–82
Pork Motel, 206
"Porkopolis." *See* Cincinnati
pork packing. *See* packing industry
*Pork People Like* (1956 film), 168
poultry, 25–27, 29, 67, 76, 154, 166, 205
power, 3, 5, 10, 220
Powers, Elmer, 131
Powhatan War (1622), 15
*Prairie Farmer*, 123, 128, 140, 187, 191

Prem (premium ham), 164
Premium Standard Farms (PSF), 27, 204, 211, 214
prices of hogs, 22, 25, 30, 130–32, 134
prime pork, 119–20
processed pork products, 177, 178, 180
Procter, William, 122
production contracts, 25–28
prosciutto, 172
protectionism, 127, 128, 134
PRRS (porcine reproductive and respiratory syndrome), 150, 151, 214
Prudence Island, 47
pseudorabies, 205
PSF. *See* Premium Standard Farms
public health, 91, 92, 94, 127, 128
purebred hogs, 54, 55, 185, 186, 197
Pynchon, John, 107

quarantines, 145, 148, 149
Querido, John, 102–3

Ræder, Ole Munch, 85
railroads, 110, *112*, 115, 118, 126, 195–96
Raper, Arthur, 160
Raritan people, 15
Rath, John, 80–81
rationing, 164
razorbacks, 9, 41, 183, 184
R. C. Farms, 103–4
red meat, 159, 166, 171
refrigeration, 126, 159
rendering, 114, 123
Rex, Samuel, 119
Rhode, Paul, 137
Rhodes, V. James, 24
ridding the guts, 160
ringing hogs, 3–5, 50, 52, 86, 193
*Ringing the Pig* (Mount painting), 3–5, *4*

283

rivers, 53, 109–10, 208, 211
Robbins, E. T., 195
Robinson, Solon, 40
Roosevelt, Franklin, 130
Rosebud Sioux Reservation, 211
Rosenberg, Gabriel, 186
Rowe, Mike, 103
Rowlandson, Mary, 16
Russell, Richard, 74

Sale, Albert Gallatin, 51, 81
Salmon, Daniel E., 141, 142, 143
salt, 78, 79, 106, 111–12, 114, 119, 174
salt pork, 2, 11, 20, 68–69, 71–74, 77–78, 124, 128, 153, 156, 159–60, 176
San Bernardino, 95, *96*, *97*, 100
Sand Livestock, 205, 212
San Francisco, 91
sanitation, 145, 146, 191, 193–95, 197–99
San Juan Island, 17–18
saturated fats, 174
sausage, 96, 114, 122, 127, 129, 155, 157, 162, 177, 180
Savage, John, 20, 155
Savage, Tacy, 20, 155
Schoepf, Johann David, 51, 76
science, 181–216
   healthy hogs, 194–98
   limits of confinement, 206–16
   open range to pasture and pen, 182–94
   overview, 181–82
   road to confinement, 198–206
*Scientific American*, 214, 215
scours, 195, 202
scrapple, 11, 70–71, 122, 156–57
Seaboard Farms, 29, 209–10
SECD (swine enteric coronavirus diseases), 150
self-feeding, 188, 193
serum treatment, 142–47
Seven Sons Farms, 215, 216
Shannon, Fred, 156

Shaw, Thomas, 128
Shay, William W., 196, 197
sheds, 191
sheep, 16, 63, 66, 68, 69, 77
Sheldon, Asa, 76
shelters, 39, 44, 57–58, *58*, *192*, 197
Shephard, Silas, 185
Sheridan, General Philip, 20
Sherman, General William T., 80
Shiloh, Battle of, 35, 36
Shirreff, Patrick, 41
Shuanghui International, 134
simultaneous technique, 143, 144, 145
*Sister Carrie* (Dreiser), 156
skillygally, 78
slaughtering, 29, 114, 115, 162–63
slavery, 3, 44, 51, 69, 71, 73–74, 81, 107, 121–22, 153, 175
Smith, Alfred, 151
Smith, John, 8
Smith, Theobald, 142
Smithfield Foods, 26, 27, 29, 134, 214, 220
smokehouses, 70, 78, 155
Snow, Edwin, 138–39
soap, 122
*Soap Locks, or Bowery Boys, The* (Calyo), *90*
soul food, 175, 176
South Carolina, 10, 14, 41, 49, 57, 74, 75
southern diet, 65–67, 74–75, 79, 153–54, 160–61, 174–77
Soviet Union, 132, 165
Spam, 2, 132, 163–65, 177–78
Spanish settlers, 7, 8, 63–64
Speer, Charlie, 214
Spicer, Jacob, 43
Spivey, A. C., 194
Staten Island, 15
*State of Bacon* (film), 179
steak, 72, 77, 154–56, 159, 166

stem cell research, 223
Stevens, Martha, 211
Stewart, John, 50
St. John, Mary, 62
Stokstad, E. L. R., 199
stray hogs, 48, 51, 86, 90
Stuart, James, 41, 46, 67, 75
Student Nonviolent Coordinating Committee (SNCC), 175
*Successful Farming*, 27, 199
sugar, 107, 121, 128, 166, 174
surface water contamination, 208, 219
Susie the Talking Sow, 196
*svinaia tushonka*, 132
Swift, 128, 129, 164
swine. *See* hogs
swine enteric coronavirus diseases (SECD), 150
swine fever, 136
*Swine Husbandry* (Coburn), 187
swine plague, 141
swine plagues, 135–51
   hog cholera, 136–49
   new diseases, 150–51
   overview, 135–36
Symonds, Major Henry C., 124

tallow, 64, 157, 174
Tamworth hogs, 172, 197
tankage, 189, 193
Tarr, Joel, 84
Taylor, Joe Gray, 153, 161
Taylor, John, 53
Telfair Plantation, 122
Tennessee, 7, 12, 18, 19, 78, 109, 121, 144
tetracycline, 208
Texas, 7, 8, 29, 32, 41, 64, 75
Texas fever, 136, 141
TGE (transmissible gastroenteritis), 150, 151
theft of hogs, 121–22, 182, 183
Thirsk, Joan, 37, 63
tinned meat, 2, 132, 163–65, 177–78

# INDEX

tobacco, 26, 36, 38, 42, 73, 106, 107
Towne, Charles Wayland, 5
transgenic pigs, 222, 223
transmissible gastroenteritis (TGE), 150, 151
transplantation, 220–23
trichina, 98–99, 127–28, 162–63, 170
Trollope, Frances, 114
Truman, Harry S., 144–45, 193
Trumbull, Jonathan, 107

*United States of Bacon* (TV show), 179
unsaturated fats, 174
urban pigs, 83–104
  confronting urban swine, 85–93
  garbage feeding, 93–100
  odor and suburban growth, 100–104
  overview, 83–85
US Army, 19, 20, 78, 79, 124
US Department of Agriculture (USDA)
  feral hogs, 31
  hog cholera, 136, 137, 141, 144, 145, 148, 149
  making bacon and white meat, 159, 162, 166
  markets, 130, 132, 133
  post-emancipation, 21
  science and swine, 184–85, 188, 195, 204
  swine plagues, 136, 151
  urban pigs, 96, 97

US Environmental Protection Agency, 103, 208, 218

vaccination, 141, 143–45, 147–49, 151, 197
Vance, Cyrus, 79
Vance, Rupert, 160
Vancouver, George, 106
van der Donck, Adriaen, 9
Van Tienhoven, Cornelius, 42
vegetable oils, 158, 167, 172, 174
Virginia, 8–10, 15, 26, 38, 41, 44, 47–48, 53, 65–67, 70, 74–75, 107
viruses, 142, 143, 144, 150, 222
vitamin B12, 199

Walker, William, 51, 53
Wallace, Henry A., 130, 131, 197
*Wallaces' Farmer and Iowa Homestead*, 130, 158–59, 186
Walsh, Lorena, 67, 72
Walsh, Margaret, 112, 126
Wampanoag people, 15
Ward, Edward, 64
Washington, DC, 86, 90
Washington, George, 42
waste disposal, 3, 6, 83, 85, 93–104, 162, 202–4, 219
water pollution, 94, 103, 204, 208–11, 213, 218–19
Waud, Alfred, *80*
Waverly Plantation, 153
weaning, 199, 202, 205
Wentworth, Edward Norris, 5
West Indies, 106, 107

whale oil, 123, 124
Whitman, Major Edmund B., 36
Wiggins, Francis, 56, 57
wild boar, 9, 31, 32, 36, 58–59
wild hogs. *See* feral hogs
Williams, David R., 122
Williams, Roger, 13
Willis, Paul, 173
Wilson, Charles Henry, 85
Wilson, James, 115
Wilson, John, 72, 152, 153
Wilson, Normal, 212
Wilson, Robert, *59*
Wilson, Woodrow, 130
Wilson and Company, 128, 164
Winthrop, John, 68
women's roles, 163, 168, 170
Woods, John, 40
working class, 88, 90, 153, 156, 159, 210, 213, 219–20
working people's food, 61–82
  Euro-American taste, 64–71
  European taste, 62–64
  making the republic of porkdom, 72–77
  overview, 61–62
  wartime pork, 78–82
World War I, 94, 129
World War II, 23, 98, 100, 131, 164, 174, 186–87
Wormley, John, 51
worms, 127, 145, 195

xenotransplantation, 220–23

yoking hogs, 4, 49–50, *49*, 52

285

CPSIA information can be obtained
at www.ICGtesting.com
Printed in the USA
LVHW072106231218
601275LV00005B/3/P

9 781946 684